BEST PRACTICES IN GEOGRAPHIC INFORMATION SYSTEMS-BASED TRANSPORTATION ASSET MANAGEMENT

January 2012

Prepared for:
Office of Planning
Federal Highway Administration
U.S. Department of Transportation

Prepared by:
Program and Organizational Performance Division
John A. Volpe National Transportation Systems
Research and Innovative Technology Administration
U.S. Department of Transportation

Report Notes and Acknowledgments

The U.S. Department of Transportation John A. Volpe National Transportation Systems Center (Volpe Center), in Cambridge, Massachusetts, prepared this report for the Federal Highway Administration's (FHWA) Office of Planning. The project team consisted of Jessica Hector-Hsu and Valarie Kniss of the Technology Innovation and Policy Division and Ben Cotton of the Transportation Planning Division. Mark Sarmiento of FHWA's Office of Planning and Chris Chang of FHWA's Office of Infrastructure provided project oversight.

The project team reviewed relevant literature and conducted an internet scan of Geographic Information Systems (GIS) and transportation asset management (TAM) applications to identify potential case studies. The project team then held interviews with contacts from public agencies listed in Appendix C. The case studies presented herein are based on these telephone discussions and supplemental materials provided by interviewees. Interview contacts reviewed draft case studies for correctness and clarity and provided additional information as needed. The Volpe Center project team thanks the staff members from the organizations across the country that contributed to this report. The time they graciously provided was fundamental in preparing the document.

CONTENTS

Executive Summary 1
1   Introduction 1
        1.1   History and Background of Transportation Asset Management 2
        1.2   Geographic Information Systems for Transportation Asset Management 4
2   Benefits and Applications 8
        2.1   Simplified Data Sharing 8
        2.2   More Accessible Data 9
        2.3   Data Entry 10
        2.4   Proactive Asset Management 10
        2.5   Informed Decision-making 12
        2.6   Public outreach 13
        2.7   Project Design 14
        2.8   Coordinated Fleet Management 15
Best Practices 16
        2.9   Institutional Best Practices 16
        2.10   Technical Best Practices 17
3   Challenges 27
        3.1   Stovepipe Organizations 27
        3.2   Difficulty Garnering or Sustaining Leadership Support 27
        3.3   Data Collection Costs 27
        3.4   Complex System Architectures 28
        3.5   Ownership Ambiguities 28
        3.6   Lack OF DATA Standards 28
        3.7   Staff Size and Time Constraints 30
        3.8   Determining How Much Data to Share 30
4   Trends and New Technologies 31
        4.1   Improved Mobile Data Collection 31
        4.2   Automatic Geotagging 32
        4.3   Cloud Computing 32
        4.4   Increased Use of Automated Analysis 33
        4.5   3-D Visualization 34
        4.6   Visualizing Future Conditions 34
        4.7   Dynamic Segmentation (DynSeg) 34
        4.8   Open Data 35
5   Ongoing Research 36
        5.1   Academic Research 36

	5.2	Collaboration among Peers	36
	5.3	Agency Staff Involved With Research and Development	37
6		Conclusions	38

Appendix A: Case Studies A-1

Appendix B: Glossary and Acronyms B-1

Appendix C: Interview Participants C-1

EXECUTIVE SUMMARY

Maps are powerful tools. They orient. They guide. They provide a sense of scale and distance. They are a universal language, accessible to people from different personal and professional backgrounds. In a transportation planning context, maps provide a flexible canvas for communicating important information such as the location of structurally deficient bridges and proposals for future improvement projects.

Geographic information systems (GIS), which are modern mapping technologies, allow transportation agencies to harness the power of maps more quickly and inexpensively than ever before. Agencies can use GIS to visualize information about their bridges, highways, and other assets. They can also use GIS to plan maintenance trips and evaluate design alternatives to manage their assets more effectively. Maps are becoming commonly used in transportation asset management (TAM) programs that help agencies to strategically invest in their infrastructure. GIS, when applied to TAM, can help agencies communicate technical information more simply, professionally, and persuasively.

This report provides background on GIS and asset management, describes how public agencies have been integrating the two, and identifies benefits and challenges to doing so. The information presented is gleaned from a literature review and interviews with several state departments of transportation (DOTs) and one county agency. The report also identifies some leading industry trends and new, innovative approaches to using GIS for TAM.

Transportation Asset Management

TAM helps transportation agencies evaluate how investment decisions today will affect the condition of physical infrastructure in the future through dedicated data, software tools, processes, and staff. The concept was introduced to the transportation industry in the early 1990s and many transportation agencies throughout the country have initiated formal TAM programs since.

Geographic Information Systems

GIS is a computer-based mapping system that allows data to be displayed on maps and analyzed based on spatial factors. GIS software packages can be used to view base maps of geographic sections (e.g., cities, counties, states, countries, etc.) with "layers" of attribute data (e.g., characteristics about those sections) over top. GIS complements traditional TAM by allowing agencies to visualize assets and asset data using maps and geospatial analysis.

Geographic Information Systems for Transportation Asset Management

Most agencies use basic GIS to view asset condition data using 2D mapping, either through display of on-screen maps or through output of printed maps. Some agencies use more advanced geospatial analysis to combine datasets for discrete geographic sections or to combine spatial data with temporal data. A further important aspect of geospatial analysis is geovisualization – the use, creation and manipulation of images, maps, diagrams, charts, 3D static and dynamic views, high resolution satellite imagery and digital globes, and their associated tabular datasets.

As agencies increasingly incorporate these more advanced tools, the ability to expand the analytical process to include identification of patterns and relationship; construction and interaction with models; and communication of findings has resulted. For example, using a spatio-temporal context to examine the differences between time of day/week/month/year can have great relevance when evaluating transportation infrastructure usage and planning. Agencies also utilize spatial optimization to generate and compare project alternatives (e.g. determining the shortest distance or minimum elevation change to reduce costs of a new roadway project). Frequently the result of this analysis process is a series of possible outcomes (scenarios) which then need to be summarized and presented for final analysis and decision-making by stakeholders, interest groups, and policy makers.

APPLICATIONS

GIS offers opportunities to streamline agency business processes through visualizing, sharing, analyzing, and monitoring asset data in ways that would not be possible with strictly numerical data. GIS, when applied to TAM, can simplify data sharing, improve data access, support proactive asset management, inform decision-making, facilitate public outreach, help project designs, and provide an interface for managing rolling stock. Many state DOTs are using GIS in their day-to-day work. Oregon DOT uses GIS for internal data sharing. Its FACS (Features, Attributes, Conditions Survey) - STIP (Statewide Transportation Improvement Program) tool is an internal, web-based mapping tool that gives all staff access to spatial (inventory) data and related asset condition information. In St. Johns County, Florida, GIS is used to plan preventive maintenance routes. Washington State DOT uses GIS to manage most of its snowplow truck fleet. Colorado DOT has developed an interactive tool, the Project Locator (ProLo) web application that allows the public to see STIP projects on a map.

BEST PRACTICES

The integration of GIS into TAM plans has both an institutional and technical component. From an **institutional perspective**, agencies have found that it is helpful to have a champion to promote the program, leadership support to use and fund it, accurate data, a consistent linear-referencing and/or asset identification system, a clear idea of the intended audience of the program, a data maintenance plan, an organization that encourages collaboration, and clear roles for funding and running the program. Agencies have also found it helpful to take a phased approach to implementation, starting with a single department or region and developing a successful program incrementally before rolling it out to the entire department. The "Challenges" section describes some of the difficulties that arise when these items are not in place.

From a **technical perspective**, there are also many ways to integrate or coordinate GIS with TAM plans and activities. Most TAM models utilize software programs that rely upon collected data, a method for storing data, and tools for analyzing data. These programs also allow staff to disseminate information. GIS relies on these same basic concepts. However, GIS programs focus on spatial data, and traditional programs used in transportation asset management tend to focus on numerical and processed data; and they each have custom software packages that are designed to focus on managing and presenting these types of data. In the past, the software packages were not necessarily compatible. Now, more asset management supporting analytical tools and database systems software are being built to directly interface with GIS software. Figure ES-1 shows the enabling technologies and applications used for integrating GIS into a TAM plan. The framework is used to describe different types of architectures in place at interviewed agencies.

CHALLENGES

While every agency interviewed for this study has made strides toward integrating GIS into a TAM program, all of them acknowledge the challenges of starting and implementing GIS within these programs, including:

- GIS is often considered a planning function while asset management is considered engineering and/or maintenance function. Organizational **stovepiping** can make it difficult to coordinate.
- Despite agency support for GIS and asset management, it can be a challenge for champions to **garner support from their executive leadership** to invest in the software and staff training.
- Most agencies do not have a **tangible way of showing return on investment** to justify the expense of implementing GIS into an already established asset management program.
- The time and **cost of collecting data** for the volume of assets that are under the jurisdiction of most state agencies can be overwhelming.
- Many GIS programs and traditional asset management programs (primarily asset-class-specific tools) have **complex system architectures**. Therefore, integration and interfaces between them can be very complicated.

- Comparing the functional fit and the cost of **in-house versus commercial software solutions** is an important step for every agency.

TRENDS AND NEW TECHNOLOGIES[1]

There are several trends in the TAM and GIS fields. Agencies on the cutting edge are using **mobile devices** to collect asset management and GIS data and **automatically geotag data** they collect with other devices. **Cloud computing** is an increasingly popular way to remotely store data and share software among many GIS and TAM users. Some agencies are already taking advantage of **GIS analysis tools** that run statistical and comparative analyses of geospatial data, viewing **3-D visualizations** of assets in virtual landscapes and showing projections of **future asset condition data** on maps. Other agencies are expected to follow in the future. **Dynamic Segmentation**, which allows more data related to single features to be stored in GIS without duplicating base data or dividing the feature into many unwieldy segments, is becoming a more common feature in GIS tools. **Open data** initiatives are allowing agencies to share more data with the public but also compelling them to question how much is practical and appropriate to share.

CONCLUSION

GIS enhances the field of transportation asset management. Maps help leaders see the extent of problems, understand the geographic impact of their decisions, and ultimately make more informed decisions. Maps can also help the public see and understand the far-reaching importance of the transportation assets they use every day. GIS enables transportation agencies to show information about their assets on maps that both technical and non-technical audiences can understand. It provides analysis tools that agencies can use to consider geographic features in the maintenance and design of their infrastructure. It also allows agencies to compare asset condition data with socio-economic, environmental, financial, and other types of data to identify relationships that they may have not considered before and make better decisions.

While state DOTs have made great strides in applying GIS to many areas of transportation, it has been a slow process in fully realizing the true potential of GIS in transportation asset management. The interviews conducted through this study suggest that there may be many reasons for this, including:

- Stovepipe organizations
- Lack of staff training and appropriate software modules (both tied to funding and leadership support)
- Difficulty in determining what is the "right" GIS data to collect to maximize benefits and therefore waiting until all rigorous analysis tools are in place before implementing GIS into asset management instead of using the existing tools and data to build a preliminary understanding of asset conditions
- Proprietary standards, different data base formats, or legacy systems that make sharing of data difficult.

Even in mature GIS and TAM programs, good data, collaborative cultures, technical standards, and leadership support are necessary keys to success.

New technology offers solutions to existing challenges and provides advanced agencies with a foretaste of even more capability yet to come. New mobile devices will make data collection less expensive and easier to do. Ongoing TAM supporting analytical tools and GIS software improvements will make data manipulation and advanced mapping easier. Open source software and industry data standards may make the market for GIS and TAM software more versatile and affordable to small agencies. These small

---

[1] Please note that although many proprietary products are described in this report in sections below, FHWA does not endorse any of these products particularly. These technologies were individually selected by each agency or DOT

changes will make it easier for agencies to use GIS in their TAM programs but will also require practitioners to follow changes in the industry and continually grow their skills.

State transportation agencies aim to provide safe, reliable, and efficient services to the users of their systems. They are also held accountable for the decisions they make and the taxpayer funding that is used to pay for maintenance and construction of the system. Taken independently, TAM and GIS activities can help agencies to provide better service and to do so in an efficient manner. Taken together, GIS can help staff communicate the findings from their TAM programs in visual way that the public can understand and leaders can use to make better decisions.

DATA MANAGEMENT

COLLECTION
- Manual / Pen & Paper
- Traditional Surveying & GPS
- GPS-enabled Handheld Device
- Maintenance Vehicle
- Aircraft (manned or unmanned)
- Satellite Imagery
- Automated Road Analyser

STORAGE
- Paper Files
- Electronic Files (.xls)
- Custom Database
- Asset Management Database
- Geodatabase

ANALYSIS
- Manual Analysis
- Custom Asset Management Analysis Tools
- Asset Management Analysis Tools
- Geospatial Analysis Tools

INFORMATION DISSEMINATION
- Manual Reports
- Custom Reports
- Asset Management Data Reports
- GIS Maps and Data Layers

Figure ES-1: Enabling Technologies for GIS and TAM

1 INTRODUCTION

This report describes how geographic information systems (GIS) applications are being used to better support transportation agencies with their asset management efforts. The practices, challenges, and lessons learned in this study are expected to help others in the transportation industry use GIS to manage their assets more effectively.

Maps are powerful tools. They orient. They guide. They provide a sense of scale and distance. They are a universal language, accessible to people from different personal and professional backgrounds. In a transportation planning context, maps provide a flexible canvas for communicating important information such as the location of structural deficient bridges and proposals for future improvement projects.

GIS allows transportation agencies to harness the power of maps more quickly and inexpensively than ever before. Agencies can use GIS to view information about their bridges, highways, and other assets (see Figure 1). They can also use GIS to plan maintenance trips and evaluate design alternatives to maintain their assets more effectively. Maps are becoming commonly used in transportation asset management (TAM) programs that help agencies to strategically invest in their infrastructure. GIS, when integrated into transportation asset management plans, can help agencies communicate technical information more simply, professionally, and persuasively.

Figure 1: GIS can be used to view transportation assets and information about them on a map, like this GIS-based Maintenance Management System used by St Johns County, Florida. Source: St. Johns County.

Best Practices in GIS-Based Transportation Asset Management

This report begins with a history of TAM and GIS, and an overview of why they are functionally integrated. Section 2, "Benefits and Applications," discusses the business activities that integrated TAM and GIS activities support at transportation organizations. This section also provides practical information about how agencies can integrate TAM and GIS, from institutional and technical perspectives. Section 3 describes some of the challenges encountered by agencies that have implemented GIS into the existing transportation asset management programs. New technologies relevant to using GIS for TAM are previewed in Section 4 and forums for ongoing research are described in Section 5. Section 6 is a conclusion and the Appendices provide supplementary material, including full case studies for interviewed agencies.

The information presented is gleaned from a literature review and interviews with several transportation organizations, including:

- Ohio Department of Transportation - Headquarters Office (Ohio DOT)
- Ohio Department of Transportation - District 2 Office (Ohio DOT)
- Washington State Department of Transportation (WSDOT)
- Oregon Department of Transportation (ODOT)
- Colorado Department of Transportation (CDOT)
- St. Johns County, Florida, Department of Public Works (St. Johns County)
- Michigan Department of Transportation (MDOT)

The report also highlights leading industry trends and new, innovative approaches to using GIS for TAM.

1.1 HISTORY AND BACKGROUND OF TRANSPORTATION ASSET MANAGEMENT

TAM has many definitions. At its most basic level, TAM is the process by which agencies decide how to invest in their physical infrastructure. Strategic TAM provides the means for agencies to look into the future and see how their financial decisions will likely affect the condition of their assets. It also helps agencies align maintenance and capital plans with agency strategic goals and measure progress towards performance goals.

TAM was introduced to the transportation industry in the early 1990s when government leaders and the public began to demand increased public sector productivity. The Government Performance and Results Act of 1993 required governmental agencies to report how and why money was spent.[2] Thousands of miles of roadway infrastructure constructed in the 1950s through the Interstate Highway System were nearing the end of their design life and funding was limited. Transportation agencies needed strategies for rehabilitating the roads and demonstrating to the public that they were being good stewards. TAM concepts, common in the private sector, were a ready solution.

The Federal Highway Association (FHWA), with leadership from the Office of Asset Management, partners with the American Association of State Highway and Transportation Officials (AASHTO), State and local departments of transportation (DOTs) along with FHWA field offices, the Transportation Research Board, and industry in encouraging the application of asset management. As defined by the AASHTO Standing Committee on Highways, Planning Subcommittee on Asset Management:

> *Transportation Asset Management is a strategic and systematic process of operating, maintaining, upgrading, and expanding physical assets effectively throughout their lifecycle. It*

---

[2] Federal Highway Administration (FHWA), "Asset Management Overview." Website. Updated 4/4/11. http://www.fhwa.dot.gov/asset/if08008/amo_03.cfm

focuses on business and engineering practices for resource allocation and utilization, with the objective of better decision making based upon quality information and well defined objectives.[3]

In following, FHWA created an Office of Asset Management (OAM) in 1999 to focus on asset performance, preservation, and longevity. OAM works closely with AASHTO and the Transportation Research Board (TRB) to provide technical assistance, education, training, research findings, and other resources to federal, state, and local transportation agencies.

Strategic TAM enables transportation agencies to communicate with stakeholders, decision-makers, and the general public through forward-looking analysis. Instead of simply declaring that the transportation infrastructure is in need of improvement, an agency is able to demonstrate, for example, the impact of changing budget levels on a system's future condition and performance. The approach allows for strategic resource allocation and unilateral communication at a time when U.S. highway infrastructure is now beyond the end of its design life, construction costs are more expensive than ever, and budgets are increasingly stretched (see Figure 2).[4]

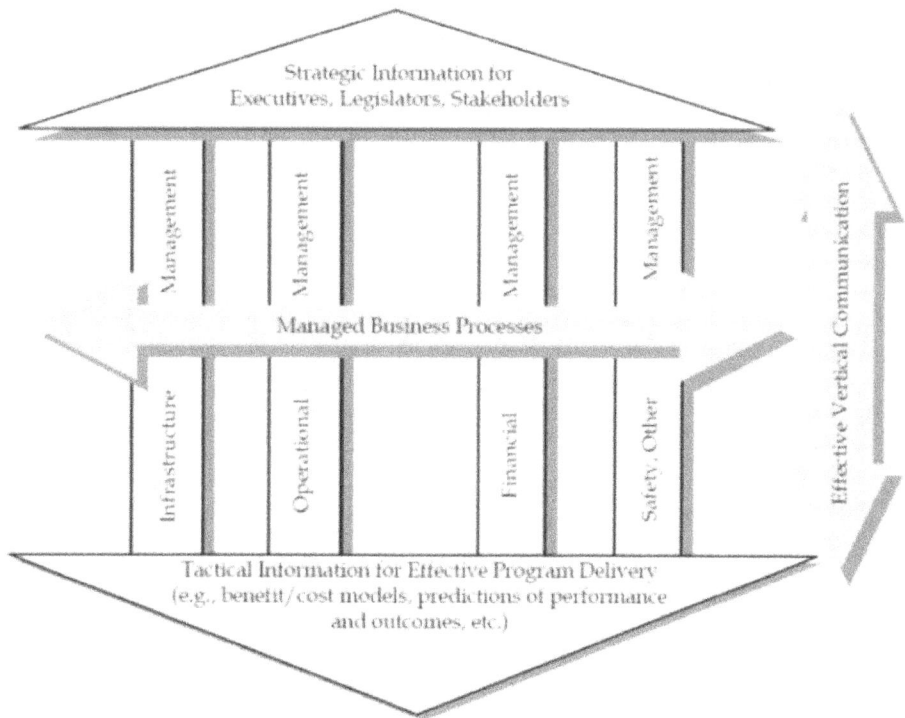

Figure 2: Transportation Asset Management is used to inform Tactical and Strategic decision-making at transportation organizations. Source: FHWA's Asset Management Primer.

Specific TAM practices vary widely by organization due to unique variations in the transportation modes, physical infrastructure types, climate, political environment, legal constraints, leadership, and historical changes that transportation agencies have faced. However, there are some high-level commonalities. As recommended in FHWA's *Asset Management Primer*, sound asset management systems are generally customer driven, mission driven, system oriented, long term in outlook, accessible, user friendly, and flexible.[5]

---

[3] This definition was developed by the AASHTO Subcommittee on Asset Management in January 2006. http://www.transportation.org/sites/scoh/docs/Motion_Trans_Asset_Management.doc
[4] FHWA, "Asset Management Overview." Notes from the Director. http://www.fhwa.dot.gov/asset/if08008/amo_01.cfm.
[5] FHWA Office of Asset Management, "Asset Management Primer." Dec 1999.

Agency TAM programs typically encompass datasets, software tools, and processes that encourage (or require) staff and leadership to consider the long-term impacts of funding decisions on the condition of their infrastructure and organizational strategic goals. Agencies collect TAM datasets that describe individual assets (or groups of assets) with unique identifiers; location information (often expressed as a milepost number along the roadway or latitudinal/longitudinal coordinates); physical attributes (e.g., materials, color, and size); age; and condition information gathered during field inspections. TAM software tools are used to analyze asset datasets – project future conditions, estimate the cost to repair or replace assets, evaluate the impacts of alternative funding levels on the condition of the asset over time, and determine whether a given level of investment will allow the assets to achieve a level of quality important to the organization. Processes and procedures guide agencies in running TAM programs that allow them to achieve their organizational goals and communicate with internal decision-makers and the public.

1.2 GEOGRAPHIC INFORMATION SYSTEMS FOR TRANSPORTATION ASSET MANAGEMENT

GIS is a computer-based system that provides a full suite of tools for the creation, management, analysis, and display of spatial data. Categories of features (e.g., roads, water bodies, jurisdictional boundaries) are grouped into "layers" and can be combined in order to create a reference basemap. Specific business data features such as poles, signs, culverts, and guardrails are also categorized and displayed on top of a referential basemap. GIS provides the typical data query functionality such as select data features/records by type, length, value, inspection date, and ownership. GIS can also query point, line, and polygon data features/records based on spatial relationships such as within an area, along a route, overlapping, within a given distance, or intersecting. Tabular queries and spatial queries can be combined in order to perform more complex analysis, and the results can be displayed on a map in order to illustrate patterns and spatial relationships. Data features can be displayed differently (color, size, pattern, width), based on various attributes (type, length, value, inspection date, ownership) in order to represent query/analysis results.

Agencies can add richness to the asset inventory/condition data by integrating with spatial components of those physical assets through GIS to view TAM data on a map either in print or on screen. Agencies are most often using these basic visualization techniques to display:

Images:Volpe Center/Google

Road Sign Example

As a specific example of GIS for TAM, consider a road sign. Basic data associated with it would be its serial number, its location along a roadway, the year it was installed, the materials it is made of, and the condition of the sign based on its last inspection.

A TAM analysis tool could be used to estimate the ideal maintenance and replacement investments over the next 10 years. It could also allow an agency to simulate the condition of the asset during the same time if it does not receive sufficient attention. It could also allow the agency to compare the needs of one sign versus another, aggregate the projected investment needs across the entire sign program, or compare the magnitude of sign program needs versus pavement program needs.

A GIS tool could be used to map the sign and analyze data associated with it. The basic spatial data associated with a sign would be its latitudinal and longitudinal coordinates and its elevation. This information, along with the physical attributes of that sign (often provided by the TAM system described above), could be stored in a geodatabase. An agency could then map that sign to assess whether it is in an effective location based on road alignment and topography. The agency could also map an entire network of signs or use a GIS-based analysis tool to identify deficiencies, like road sections that lack signs.

- **Asset inventory data**, visual cues that an asset or a group of assets is located in a geographic expanse;

- **Current asset attributes** such as the age, model, or condition of the assets;

- **Future asset attributes**, projections (usually developed outside of GIS) of what the condition or form of an asset will be at a date in the future;

Within basic visualization, there are several viewing methods for the maps. Some agencies produce GIS maps that are **static**, prepared in advance, presented as is, and not alterable by the audience. Some also have **interactive** maps, available through a user interface that allows the user to tailor the information and view.

Agencies have found the maps to be useful for internal and external purposes. **Internal** maps are prepared to inform and educate agency staff. **External** maps are designed for the public to view information that may be of interest (such as new construction projects).

In addition to basic visualization, GIS can be used to perform complex geospatial analyses. Through those analyses, practitioners can view, compare, and contrast datasets for specific geographic sections. They can also compare project solutions and operating scenarios based on spatial criteria or temporal data. . While the majority of agencies are still using GIS for basic visualization techniques for TAM, some agencies are using geospatial analysis for:

- **Location-based data queries.** This capability allows agency staff to isolate assets that meet certain spatial criteria. For example, a GIS analysis tool can be used to identify all roads within a state that are located within 100 feet of a wetland. Staff then use this information to focus their stormwater runoff management efforts. Thus, GIS is effectively used to support project-level asset management decisions.

- **Statistical analysis of two or more datasets for a geographic area.** Statistical analysis tools help practitioners determine whether there is correlation between different sets in a geographic location. For example, staff can use GIS analysis to determine whether there is a relationship between income and pavement condition in different neighborhoods of a city. If an agency determines that there is an apparent relationship (i.e., lower income neighborhoods have poor pavement conditions), it may choose to address the disparity.

- **Comparing project alternatives based on defined spatial criteria or temporal data.** GIS analysis tools help agencies identify the solutions to problems based on spatial or temporal data. For example, staff can use GIS to identify the least-cost path from point A to point B based on distance, elevation change, and sensitive ecosystems through which alternative routes might pass. Using a spatio-temporal context to examine the differences between time of day/week/month/year can have great relevance when evaluating transportation infrastructure usage and planning. Frequently the result of this analysis process is a series of possible outcomes (scenarios) which are presented to stakeholders, interest groups, and policy makers for decision-making purposes.

- **Asset optimization and cross-asset optimization.** GIS is used to strategize and optimize individual asset life cycles as well as to better coordinate the timing and scheduling of cross-asset activities. For example, some agencies are recognizing the ability of GIS to look at strategic scheduling of maintenance across asset classes to avoid utility repairs on newly resurfaced roadways and other costly non-coordinated maintenance activities. The ultimate goal is to find a solution that considers all asset classes simultaneously, considers existing conditions, predicts future conditions, and allows evaluation of different budget scenarios and resource constraints. This is highlighted through the concept of "global optimum" versus "local optima". An example of a globally suboptimum decision is when a critical maintenance project is not implemented due to budget constraints while a less critical project, funded by a different budget based on asset category, goes forward. Within their respective budget realms, the prioritization may have been optimal in a narrow, short-term sense, but overall, the maintenance program is suboptimal. The experience of agencies in this type of optimization using GIS has been limited but is progressing.

- **Asset data collection, inspection, and maintenance.** GIS is of particularly high value in field applications. A handheld device with a GIS map, a GPS receiver, and a digital camera provides a powerful platform for collecting, inspecting, and maintaining asset features.

As an example of location-based analysis, St. Johns County, Florida looks at the growth of its asset inventory in relation to their maintenance district boundaries. Then, based on the total number of assets within each district and its associated maintenance budget for the assets, the County will redraw boundaries to ensure its maintenance tasks are evenly distributed. The County is also using spatial optimization that works directly with GIS data to group maintenance activities into more spatially logical projects for implementation.

Table 1 shows ways that interviewed agencies are using GIS as a basic visualization tool and for geospatial analysis in their TAM programs.

Table 1: Ways in which Interviewed State DOTs have used GIS as a visualization tool in Transportation Asset Management Programs

	Basic Visualization – Data			Basic Visualization – Audiences and Viewing Methods				Geospatial Analysis		
	Inventory	Current Attributes	Future Attributes	Internal Static Maps	External Static Maps	Internal Interactive Maps	External Interactive Maps	Location-Based Queries	Statistical Analysis	Optimization of Solutions
Ohio DOT	✓	✓		✓	✓	✓	✓			
Oregon DOT	✓	✓		✓	✓	✓	✓			
Michigan DOT	✓	✓	✓	✓	✓	✓	✓	✓	✓	✓
Colorado DOT	✓	✓		✓	✓	✓				
St. Johns County, FL	✓	✓		✓	✓	✓	✓	✓		✓
Washington State DOT	✓	✓		✓	✓	✓		✓		

2 BENEFITS AND APPLICATIONS

The main reason for integrating TAM and GIS is that GIS offers opportunities to streamline agency business processes through visualizing, sharing, analyzing, and monitoring asset data in ways that are not possible with traditional tabular data, graphs and charts. These opportunities help simplify data sharing across groups, improve data access for all employees, reduce the cost of proactive TAM, inform internal decision-making processes, encourage stakeholder involvement, and improve project designs.

2.1 SIMPLIFIED DATA SHARING

GIS enables transportation agencies to easily share spatial data, digitally. Anyone with compatible software can view a GIS map on screen and use basic interactive features (such as turning layers on and off). Most importantly, digital maps can be updated and distributed quickly.

Oregon DOT (ODOT) uses the FACS-STIP (Features, Attributes, Conditions Survey- Statewide Transportation Improvement Program) tool, an internal web-based mapping tool, which was developed to help employees by improving transportation data availability. The goal of FACS-STIP is to make asset data available for scoping and design, improving the overall project delivery and asset management processes. This tool helps users make informed decisions by making basic inventory and condition information available for various highway assets from numerous databases. Prior to FACS-STIP, there was no "one-stop-shopping" for all statewide collected data in Oregon. ODOT is using FACS-STIP to assist with planning, STIP development, and scoping, construction, and maintenance by sharing data across regions and departments. The tool also communicates new or updated asset information with one easy-to-use application.

Ohio DOT is currently working on developing an enterprise web-based geographic information system application ("WebGIS") to more effectively distribute and communicate information related to transportation assets. The application will allow users to display map views of assets and their attributes, pan, zoom, measure distances, turn layers on and off, perform advanced data queries, and easily update data layers as needed. Integration of data sources is a key functionality of WebGIS. The application will employ user-friendly technology and integrate with various Ohio DOT existing systems, such as Ohio DOT's linear referencing system (the Base Transportation Referencing System (BTRS) which consolidates the department's various referencing systems) and

Oregon DOT's FACS-STIP Tool

Users of Oregon DOT's (ODOT's) FACS-STIP (Features, Attributes, Conditions Survey- Statewide Transportation Improvement Program) Tool can use the simple user interface to select any combination of layers:
- Base layers – such as road network, aerial imagery, regions and districts, functional class network, shaded relief, freight, and counties and cities
- Supporting layers – such as traffic flows, traffic projections, crash data, STIP projects, and number of lanes
- Project layers – such as bridge projects, pavement projects, safety project lists, and project comments

Users are able to zoom to a particular location based on project, highway mile point, latitude/longitude coordinates, and ODOT region. The system does not require any specialized training in GIS and is accessible from any computer with internet access. This tool uses ArcGIS Server software to provide information over a variety of base map options (streets, imagery, topographical and light grey). ODOT is in the process of incorporating significant additional feature data into these maps.

existing applications, such as the video-log viewer application (PathWeb) and the straight-line diagram web application. Google Streetview will also be incorporated with online basemap layers, including aerial imagery, topography, terrain, and shaded relief.

As a final example of simplified data sharing, Washington State DOT (WSDOT) has created a single source of data for all business units. The Roadside Features Inventory Program (RFIP) is a corporate program used for collecting, storing, and reporting roadside features such as guardrails, culverts, signs, and other features from all WSDOT regions. Previously, individual business units at WSDOT collected similar information independently of one another. This caused duplicate efforts, additional expense, and data entry in non-standard data formats that prevented data sharing. WSDOT's primary objective for collecting information on roadside assets is for highway safety analysis, but it has also provided essential data for the development of it transportation asset management program priorities. Thus, WSDOT has been able to integrate roadside physical asset data to support the agency in efforts beyond asset management through GIS.

2.2 MORE ACCESSIBLE DATA

In addition to making asset condition data easy to share, GIS can also make it understandable to staff with a range of backgrounds and skills sets. Asset condition data is typically collected and reviewed in specialized technical formats (e.g., inspection reports, equipment readings, electronic imaging records). TAM supporting analytical tools process numerical data and produce reports with textual or graphical representations of numerical data (e.g. charts and graphs). Individuals with the background and training to interpret the datasets (e.g., engineers and planners) can use them. However, they are less accessible to non-engineers within an organization and require effort to translate before sharing with the public.

GIS, in contrast, offers a "common language" for viewing asset data that transcends these challenges. Most people are familiar with reading maps and can evaluate visual data much more quickly than data presented in other forms. GIS does not replace all other types of data analysis, but it provides an accessible means for sharing data and communicating. Many agencies and commercial-off-the-shelf (COTS) TAM supporting analytical tools (e.g. management system tools) vendors have already been taking advantage of GIS to view TAM data in addition to purely numerical reporting.

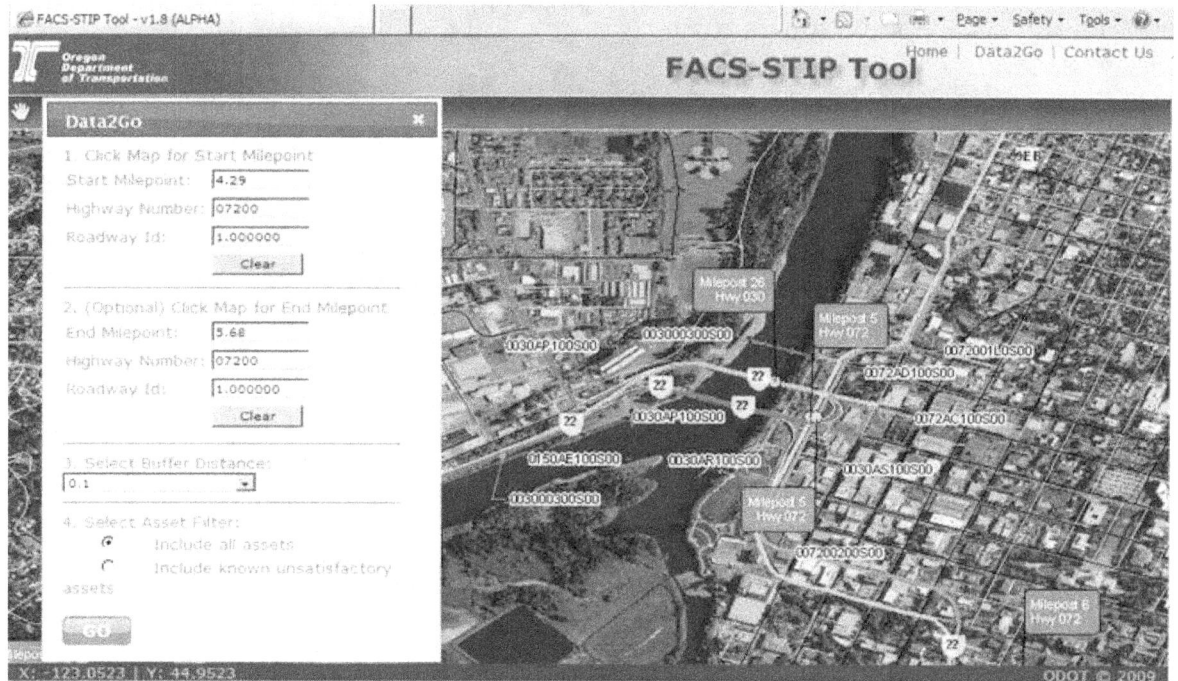

Figure 3: ODOT's Data2Go Tool allows users to select and view data from a map interface. Source: ODOT.

Best Practices in GIS-Based Transportation Asset Management

ODOT uses Asset Data2Go as part of their FACS-STIP tool (see Figure 3). Asset Data2Go allows the user to query, investigate, and export data for a particular area of interest. The area of interest can be selected by using the map interface. Once the area has been selected, the user can view and investigate the data query results for assets within it. Assets can be filtered to generate reports that can be viewed on screen or exported to Excel for further analysis. Asset Data2Go provides data in an easy-to-use format.

2.3 DATA ENTRY

Some agencies have gone beyond viewing information in GIS and have built interfaces to enter and manipulate data through web-based GIS viewers. These interfaces allow staff to enter data visually, in a manner that is more intuitive than text-only methods. Theses interfaces also enable mobile data entry. Devices and applications with GPS locators have been developed to allow staff to check out data from the office, update it in the field from a mobile device, and then check it back in with additions incorporated in new GIS layers.

Colorado DOT (CDOT) uses a GIS interface called SAP Project Manager (see Figure 4 on next page). It is a simple, map-based web interface that allows CDOT's project and corridor managers to create, modify, and review information about a project's location and other features/attributes associated with that location from the GIS interface. Through this interface, they can also track the business and financial workflows related to the project.

A unique feature of ODOT's FACS-STIP tool is the Map Commenting tool that allows users to select any point or line on the map and insert a comment for other users to view. Users can also click on a link to view the RSS feeds (Really Simple Syndication feeds, short web feeds designed to notify audiences of updates to frequently-changed web items) of submitted comments. From the RSS feed, users can see who authored the comment, the region from which the comment came, the type of comment (project related, asset related), and more specific details about the comments uploaded through Excel. There is also a map button that allows users to zoom to the location of the comment.

2.4 PROACTIVE ASSET MANAGEMENT

Proactive asset maintenance and facility improvements are essential in good TAM practice that adheres to performance management principles. GIS helps make proactive maintenance more efficient and cost-effective by helping asset managers

Figure 4: A GIS map that was developed by St. Johns County, Florida, to represent pavement improvements as part of a community redevelopment project. Source: St. Johns County.

visually plan routes and see all of the assets that can be cared for along them. St. Johns County in Florida believes that department operates much more efficiently now that it uses its GIS-based TAM

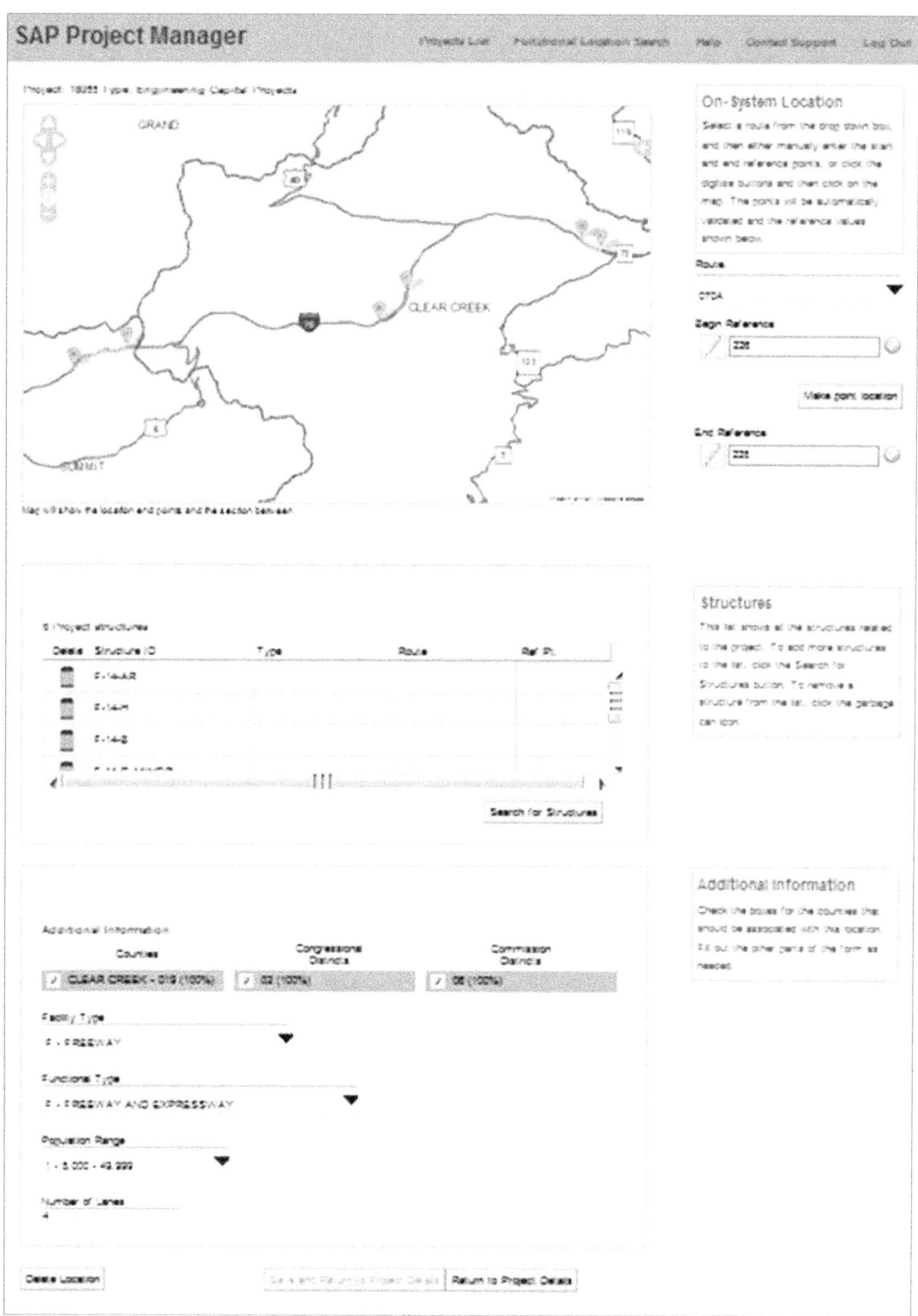

Figure 4: CDOT's GIS interface is a simple map-based web interface that allows project and corridor managers to create, modify, and review information about a project's location from the GIS interface. Users can also track the business and financial workflows related to the project. Source: CDOT.

> **GIS for Proactive Asset Management in St. John's County, Florida**
>
> In 2005, the Public Works Department of St. John's County, Florida, conducted a full review of its business operations, including TAM practices and maintenance operations. The County found that 75 percent of maintenance being performed was responsive and only 25 percent was proactive. In order to keep up with population growth and customer demands, the Department wanted to reverse this trend and work more efficiently.
>
> As a result of this review, the Public Works Department began implementing a new computerized maintenance management system (CMMS) based on GIS that would be used for both financial management and TAM. The Department's main goals were to track service requests and work orders, effectively locate the assets associated with them, and account for the costs of resolving the issues.
>
> Daily maintenance is now based on a proactive, methodical approach in which the County is divided into four work zones. Each zone has a designated team that is responsible for all routine maintenance within that zone. Work is scheduled two weeks ahead and mapped out along segments of roadway in an area. GIS enables query of all assets that fall within the designated area. The County then plans its work as a "sweeping" motion, where all maintenance-related tasks, including culvert clean-out, washing/straightening signs, repairing reflectivity, etc that fall within a mapped area are carried out during the trip. The sweeping motion saves on crew and equipment relocation costs. The County is also able to systematically track how often maintenance is completed for each asset.
>
> With this system in place, maintenance is now 80 percent proactive and 20 percent reactive. The county can limit employee overtime and better predict maintenance costs for annual budgeting purposes. In the first full year of implementation (2006–2007), the system helped increase productivity by more than 13.3 percent, saving St. John's County over $650,000.

system to prepare a predetermined schedule and path for preventative maintenance rather than waiting for assets to fail and jumping from repair to repair.

WSDOT employees also see opportunities to use the data collected from the agency's GIS-based Maintenance Operations to automate materials purchases and evaluate treatment methods. Maintenance Operations allows the maintenance staff to systematically track roadway treatment materials (e.g., sand, deicers) usage to 97 percent accuracy, which allows them to accurately monitor stock levels and automate materials requests when they run low. As more data is collected each year, WSDOT plans to use the system to measure the effectiveness of different treatments on various road conditions.

2.5 INFORMED DECISION-MAKING

Every agency interviewed reported that GIS helped improve communication between TAM practitioners and agency decision-makers. Visual data can be easier to understand and interpret than numerical data; meaning that those who are not involved with day-to-day TAM efforts can easily understand information and make informed TAM-related decisions.

Ohio DOT has been managing assets for many years and the recent availability of mapping technology has allowed the agency to display asset data visually more easily than ever before. While the agency does not often work with spatial analysis, the ability to make simple maps has proven quite successful as it pertains to influencing decision makers and agency executives. According to staff, a simple color-coded map can make a strong impact on those in decision-making positions.

Agencies have also been using GIS software to develop high-quality presentations and to compile data in a portable document format that is easily distributed. This information can be used for public presentations. For example, St. Johns County compiled various alternative scenarios when planning the SR9B Extension and prepared four different GIS maps that were presented to the public and posted to the Department of Public Works website. Shown on Figure 5 is a graphic that was developed to represent pavement improvements as part of a community redevelopment project.

GIS in TAM can also help support decision-making in court. WSDOT found that the Maintenance

Operations system allows them to provide data-driven responses to accident and litigation claims that the agency did not provide the required level of service when incidences occurred. This use is unexpected but has proven valuable.

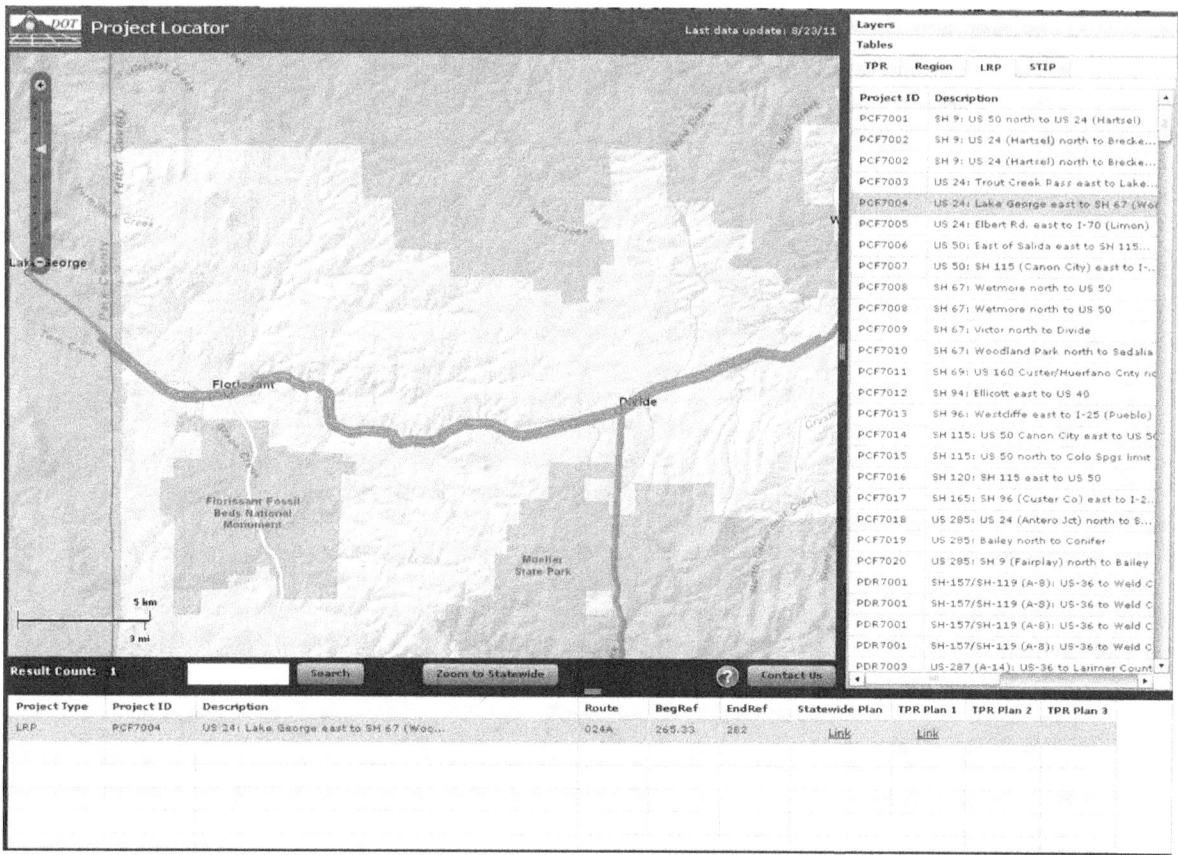

Figure 5: CDOT's ProLo web application is an interactive, public-facing, web-based map application that contains detailed information about the 2035 Statewide Transportation Plan corridors and STIP projects. Source: CDOT.

2.6 PUBLIC OUTREACH

Several agencies have created public-facing web-based mapping tools. They allow members of the public who are interested in specific construction projects or concerned about agency performance to view data on a map. Most agencies have developed websites that distribute data in a GIS format that is downloadable.

ODOT recently rolled out a new version of TransGIS 2.0 that will continue to be developed into a public facing mapping tool designed for users of many skill levels. Incorporating much of the data already collected and housed by FACS-STIP, the tool offers an interactive format that allows users to manipulate an on-screen map by turning layers on and off.

CDOT has developed a website that allows the public to view STIP projects on a map. The Project Locator (ProLo) web application (Figure 6) is a public-facing geographic web-based application that contains detailed information about the 2035 Statewide Transportation Plan corridors and STIP projects. An interactive map is available for users to search and locate corridors and projects throughout Colorado. Much of the state's GIS data is available for download to the public through the CDOT website.

MDOT is required to report annually on its transportation system conditions to the State Legislature. GIS-based graphics are developed to show the location of highway, airport, and transit assets and their conditions for these annual reports. Maps are also developed for published reports to the public, such as the "Transportation System Performance Report" shown in Figure 7. MDOT staff has found these map-based graphics to be effective in communicating the need for investment in the transportation system.

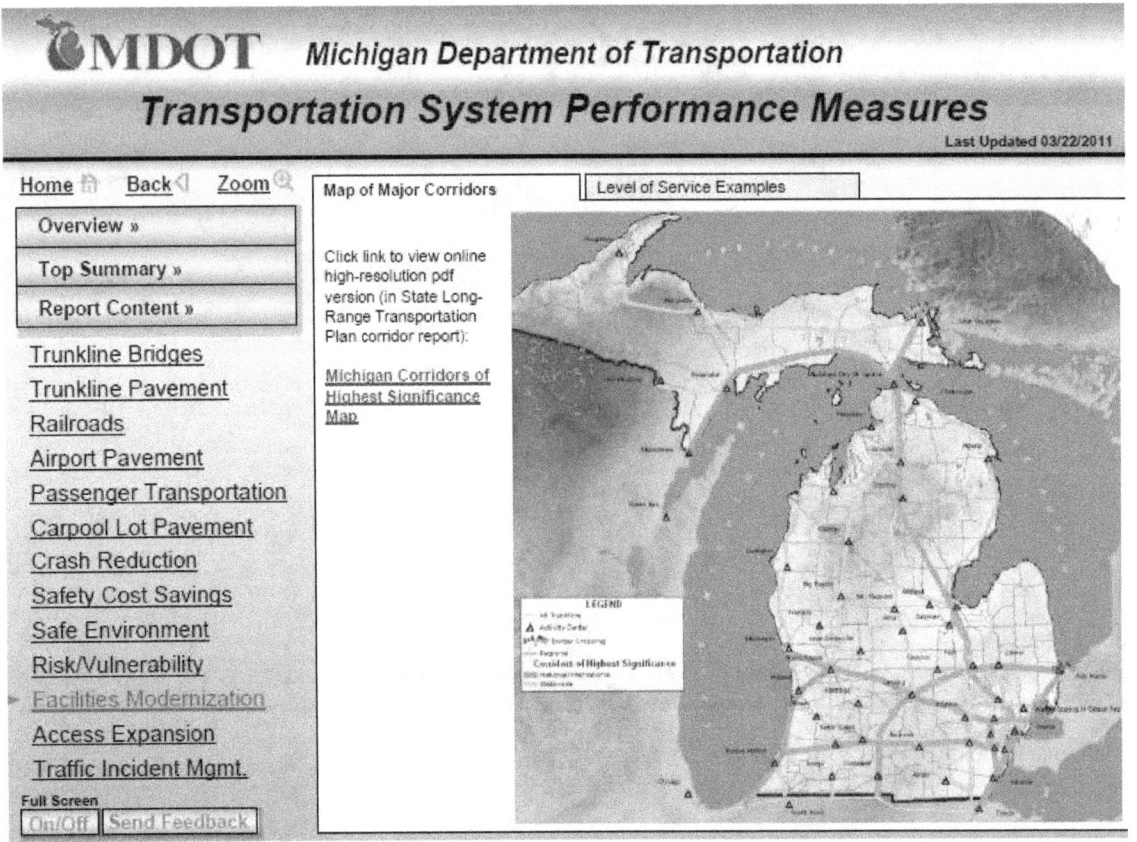

Figure 6: GIS is used by the Michigan DOT to develop graphics for public reports on asset management, such as the web-based "Transportation System Performance Report." Source: www.mi.gov/documents/mdot/MDOT-Performance_Measures_Report_289930_7.pdf.

2.7 PROJECT DESIGN

Some agencies use GIS to help scope and design projects through the use of GIS-based decision support tools. One use of GIS models is in assessing possible alternatives of various design scenarios. GIS may help agencies improve design projects if they take into account asset data early in the decision-making process. Agencies are recognizing the ability of GIS to look at strategic scheduling of maintenance across asset classes to avoid utility repairs on newly resurfaced roadways and other costly non-coordinated project activities. The ultimate goal is to find a solution that considers all asset classes simultaneously, considers existing conditions, predicts future conditions, and allows evaluation of different budget scenarios and resource constraints. This is highlighted through the concept of "global optimum" versus "local optima". An example of a globally suboptimum decision is when a critical maintenance project is not implemented due to budget constraints while a less critical project, funded by a different budget based on asset category, goes forward. Within their respective budget realms, the prioritization may have been optimal in a narrow, short-term sense, but overall, the maintenance program is suboptimal. The experience of agencies in this type of optimization using GIS has been limited but is progressing.

GIS tools can also take into account the interconnections between the various aspects of planning. Beyond asset management, GIS models can estimate the projects land use and environmental impacts. For example, if a town is trying to redesign an intersection, it could be helpful for the town to consider the characteristics of the community beyond the intersection and immediate streets entering into it, such as schools or neighborhoods that may affect usage of the intersection.

2.8 COORDINATED FLEET MANAGEMENT

GIS can be used to manage internal resources such as fleet vehicles that support maintaining physical assets on highways. Map-based interfaces are ideal for monitoring location and data feeds from maintenance vehicles, plow trucks, portable dynamic message signs, and movable Intelligent Transportation Systems (ITS) devices. WSDOT uses GIS to manage 250 trucks from its snowplow fleet (see textbox below and Appendix A).

> **WSDOT's Maintenance Operations Tool for Fleet Management and Efficient Operations**
>
> WSDOT's "Maintenance Operations" (formerly "Winter Operations") system collects real-time data from WSDOT maintenance vehicles across the state and displays the information on a browser-based map interface. The system enables employees from WSDOT's six geographic regions, who are responsible for maintaining regional roads, and the WSDOT headquarters Maintenance Operations group, who are responsible for overseeing the regions and budgeting, to manage plow operations and road conditions from a map. For any state road, staff can view up-to-date surface condition data and maintenance activity such as the material type being used for deicing and the number of plows being deployed to an area.
>
> Data is collected from 250 plow trucks outfitted with sensors that automatically collect a wide range of information regarding the truck's activities and environmental conditions. Trucks are able to capture road and air temperature, material output rates and locations, vehicle location and speed, vehicle direction of travel, driver identification number, and plow and spreader activity. WSDOT employees can view the data on a browser-based map, zoom to the appropriate level of detail, select active trucks or road segments, and view color-coded roadway conditions (ice, snow, bare and wet, bare and dry, slush, etc.) and material used on the roadway (solid chemical, liquid chemical, sand, or other) at any point in time.
>
> The most common use of the system is to support staff with shift changes in the winter – incoming employees can see where trucks have been and where work still needs to be done. The color-coded map view of all state roadways also allows maintenance managers to quickly identify hazardous conditions and intuitively assess the geographic extent and relative severity. They can then manage plow routes and deploy trucks across regions as needed to address statewide conditions, localized weather events, and weather forecasts. Finally, WSDOT management can use the application to replay and debrief storm response activities for internal learning and in response to questions from the state transportation secretary or other officials.

BEST PRACTICES

Before agencies can see the benefits of using GIS in TAM as described in the previous section, they must determine how to integrate the two programs. The integration of GIS and TAM has both institutional and technical components.

2.9 INSTITUTIONAL BEST PRACTICES

From an institutional perspective, agencies have found that it is helpful to have:

> **Ohio DOT's Phased Experience with Custom GIS and TAM Software**
>
> Ohio DOT has many new ideas about where to take its GIS and asset management programs. Rolling out new products over the entire state can be difficult, expensive, and risky. To better manage its resources and program outcomes, the department looks to its districts, counties, and municipalities to develop and test ideas and products before adopting a new program statewide.
>
> Districts have been developing customized in-house applications for the State. Ohio DOT District 2 has been the forerunner in developing several asset management applications. Ohio DOT District 1 has assisted with the implementation of these technologies. For example, District 2 is currently working on applications in advanced remote asset collection, such as the ODOT Video Log that uses a van to collect data and projects the GPS coordinates into the image as it is collected. After evaluating several vendors with video log solutions for asset extraction, the DOT determined that all were too expensive and none fit their requirements exactly. This resulted in the District building its own asset capture software. The software was developed for scalability, flexibility, and independence from any vendor. IT also leverages existing resources and investments.

- A champion, an individual or group of individuals at the organization, who can promote the cause, provide continuity, and push for change within the organization.
- Leadership support, particularly from managers with the authority to enact change and make funding decisions.
- A comprehensive and consistent data inventory for assets to be included in the effort. Agencies do not necessarily need exhaustive data for every type of asset so the selection should be strategic and purposeful.
- A consistent linear referencing system and/or asset identification system so that assets are uniquely identified and clearly located regardless of who collects and enters data.
- An idea of who the intended audiences are and what types of information they can/need to see along the way so that the right staff skills and tools are available to fulfill those requirements.
- A data maintenance plan and systems for managing updates, especially for public facing tools.
- An implementation plan that starts small, with one group or one region, and expands later into other groups or other regions across the organization once the system reaches its maturity.
- An organizational structure that encourages data sharing and cooperation by placing GIS and asset management groups in a complementary location within the hierarchy or locating some skilled GIS staff in the asset management group.
- Clear roles that define data policies and procedures and enable effective stewardship of transportation asset data.

Section 3, "Challenges," describes some of the difficulties that arise when these items are not in place.

Beyond these core values, agencies vary in their institutional implementation of GIS and TAM. Some agencies (such as MDOT, WSDOT) operate a centralized model, where core GIS and TAM activity occurs at the DOT headquarters with field or regional offices providing data collection and support. Other agencies (such as Ohio DOT) operate a decentralized model, where core GIS, and to some extent TAM, activities are initiated at the regional level with resulting information being fed to the headquarters for higher level purposes.

2.10 TECHNICAL BEST PRACTICES

From a technical perspective, there are also many ways to integrate or coordinate GIS with TAM plans and activities. Most TAM models utilize software programs that rely upon collected data, a method for storing data, and tools for analyzing data. These programs also allow staff to disseminate information. GIS relies on these same basic concepts. However, GIS programs focus on spatial data, and traditional programs used in transportation asset management tend to focus on numerical and processed data; and they each have custom software packages that are designed to focus on managing and presenting these types of data. In the past, the software packages were not necessarily compatible. Now, more asset management supporting analytical tools and database systems software are being built to directly interface with GIS software.

Figure 8 provides a high-level view of enabling technologies for integrating GIS into TAM models. Data is collected from the field using a variety of manual and automated methods; next, it is stored at the agency; then it is analyzed using custom or COTS tools; and, finally, it is viewed and/or disseminated. The framework is used in the context of this report to describe some of the functional software architectures in place at agencies that integrate GIS and TAM tools. It will also be used to illustrate potential opportunities for eliminating redundancies and streamlining business processes.

2.10.1 Data Collection Methods and Tools

The beginning and the foundation of any TAM or GIS program is quality data. Data collected from the field is the basis on which analysis and presentations can be performed. Data collection is also the most expensive part of any GIS or TAM program so it is also the most important to maximize. The data collection column in Figure 8 identifies the many ways that asset data can be procured, including:

- Many agencies still use **manual methods** to gather asset characteristics and condition data like "windshield surveys" for pavement inspections. Pen-and-paper or electronic forms are often used to document the results of these manual methods.

- Surveying can be used to collect data about the location and size of assets in electronic formats that can be converted into formats readable by GIS databases. Some modern surveying equipment, such as the "total station," employs GPS technology and allows agencies to collect data with very precise location information and enter it directly into GIS databases.

- **Handheld computers, smartphones, and tablet computers** have been most commonly used to record data that is collected manually or with other equipment by inspectors. However, they are becoming tools for data collection as well. Some innovative agencies are working to build mobile devices that can collect multiple linear features simultaneously and allow data updates in the field such as moving linear feature locations or modifying attributes.

> **CDOT's Storm Water Inventory Tool**
>
> **CDOT** has developed an application called the Storm Water Inventory Tool (SWIT) that uses Esri ArcGIS mobile and operates on a touchscreen tablet with a recreational Garmin GPS device. It is a numeric/alpha coding system designed to automate and standardize surveying and aerial mapping. A code is assigned when any topographic, drainage, utility, or aerial survey data is collected. The data is then exported into a standard GPS data format. CDOT has found that a $100 recreational device can run the application as well as a $2,000 professional grade GPS device. As a result, staff can order and deploy new devices more quickly and cost effectively.

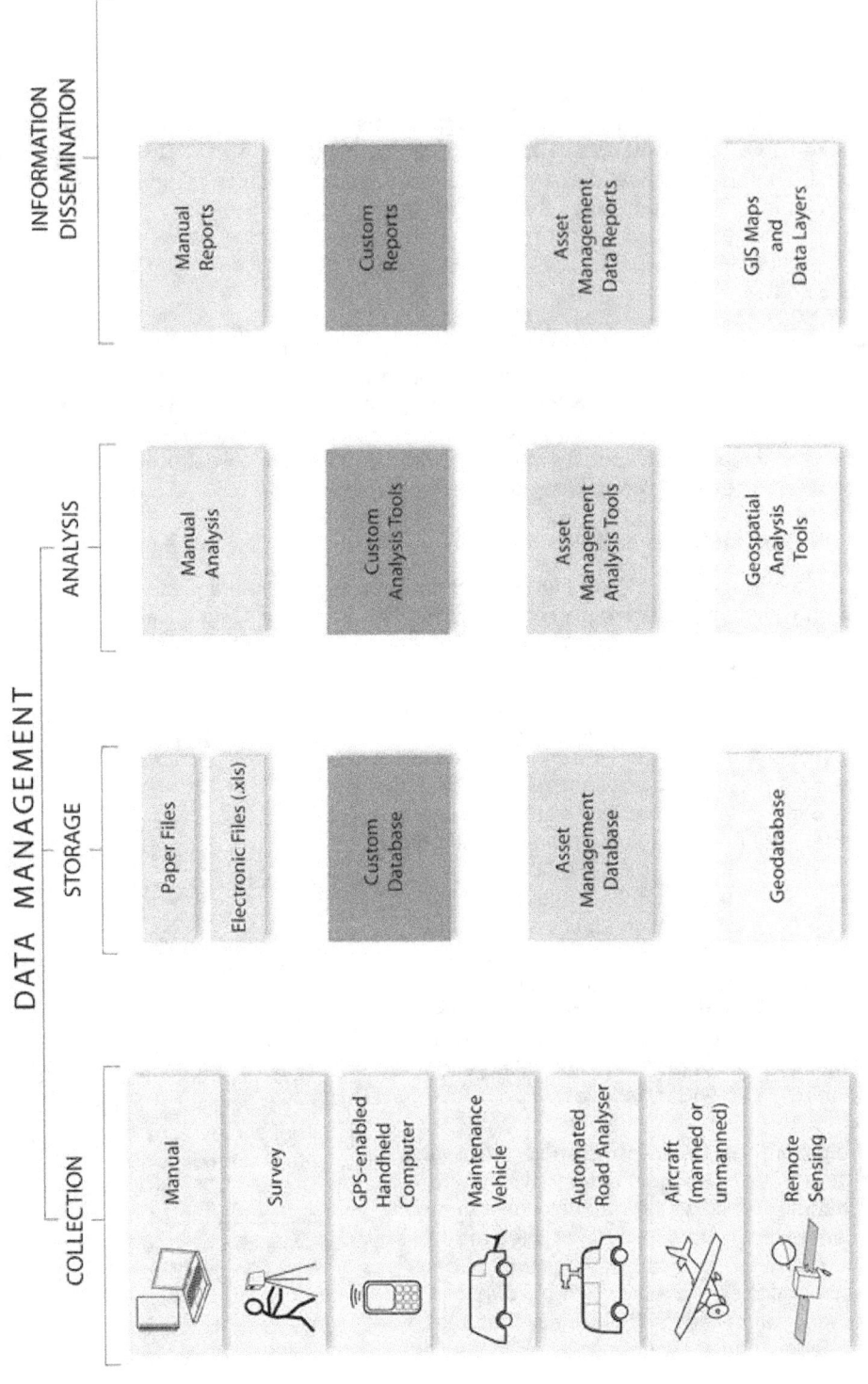

Figure 7: Enabling Technologies for GIS and TAM

Handheld computers with internal GPS offer many of the features available in a traditional survey, with less equipment. They are not as accurate as total stations, though, so they are not practical when precise measurements are needed. See Section 4.1 for more information about the future of mobile data collection.

- **Sensors and cameras** on maintenance vehicles can be used to automatically collect real-time, continuous maintenance and asset data. WSDOT uses sensors to monitor the use of deicing materials.

- Many states are now using **automatic road analyzers**, which are vehicles equipped to collect highway data while traveling at highway speeds. The vehicle collects data related to pavement condition and ride quality in addition to continuous 360-degree imagery. Several different companies have developed vehicles capable of automated data collection.

- **Aircraft surveillance** is used to gather asset data when an aerial perspective is required. This method covers a lot of ground in a short time span, but fly-overs can be expensive. Some agencies are investigating the potential of unmanned aircraft, or drones, which have become more commercially affordable. The flight plan of a drone can be pre-programmed or controlled during flight. Video is the main form of data that can be collected at this time, but there is the potential for other types of extensive automated data collection. Ohio DOT (see Figure 8) recently purchased a mini-drone with a high-resolution camera that is set to fly next spring.

Sometimes, data collection efforts and the resulting data can serve the needs of both TAM and GIS programs; other times, data collected from one type of equipment is useful for GIS but not TAM and vice-versa. Interviewed agencies described data that came from many different sources into their GIS programs; each of them recognize the financial and technical benefits of combining data collection efforts, presumably by understanding the technologies available to them and how they can be used for more than one purpose. Table 2 shows how the data collection methods can be used to collect the types of data needed for TAM and/or GIS.

Table 2: Types of GIS/TAM data that can be collected via different collection methods

	Rough Location	Precise Location	Attributes	Condition	Photographs and/or video
Manual Data	✓		✓	✓	
Survey		✓	o		
GPS-Enabled Handheld Devices	o	o	o *	o *	✓
GPS-Enabled/Sensor-Equipped Maintenance Vehicle	o	o	✓	✓	✓
Automated Asset Data Collection Vehicles	o	o	o	✓	✓
Aircraft	o	o	o	o	✓
Satellite	o	o	o	o	✓

✓ = Data can be collected; o = Data can be collected by some versions
* = Data is usually collected by another means (manually by inspector or using another device) and stored in the handheld computer.

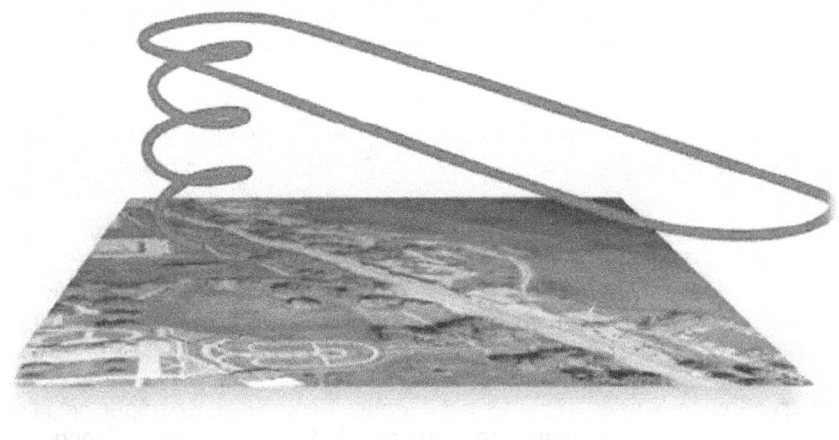

Figure 8: Ohio DOT is exploring using aircraft surveillance to collect geo-referenced asset data. This graphic is used by staff to show the potential flight path of an unmanned drone and photographs it could take. Source: Ohio DOT.

Some agencies have developed tools that make it possible for data to be collected by engineers in the office through a video log rather than physically traveling along the roadway.

Oregon DOT initiated the earthmine Pilot project (see Figure 10) to collect location, inventory, and measurement information for multiple assets and features from video and photography. earthmine for ArcGIS is a mobile interactive mapping system that displays high-resolution, road-level views alongside the traditional 2D map view. Assets can be identified, or "tagged," in the views and entered into the geodatabase with a very accurate location. Using this tool, staff are able to take measurements; view and edit attribute data; and populate the geodatabase by pointing and clicking at their desktop. The scope of the project was approximately 100 centerline miles. So far, ODOT has tagged 14 different asset categories such as retaining walls, roadbed centerline, sidewalks, sign installation, traffic barriers, and traffic structures. ODOT has found that efficiencies are gained from using such a tool: the ability for work to be completed during inclement weather conditions and improvements to the reliability of integrated data building on one coordinate data set.

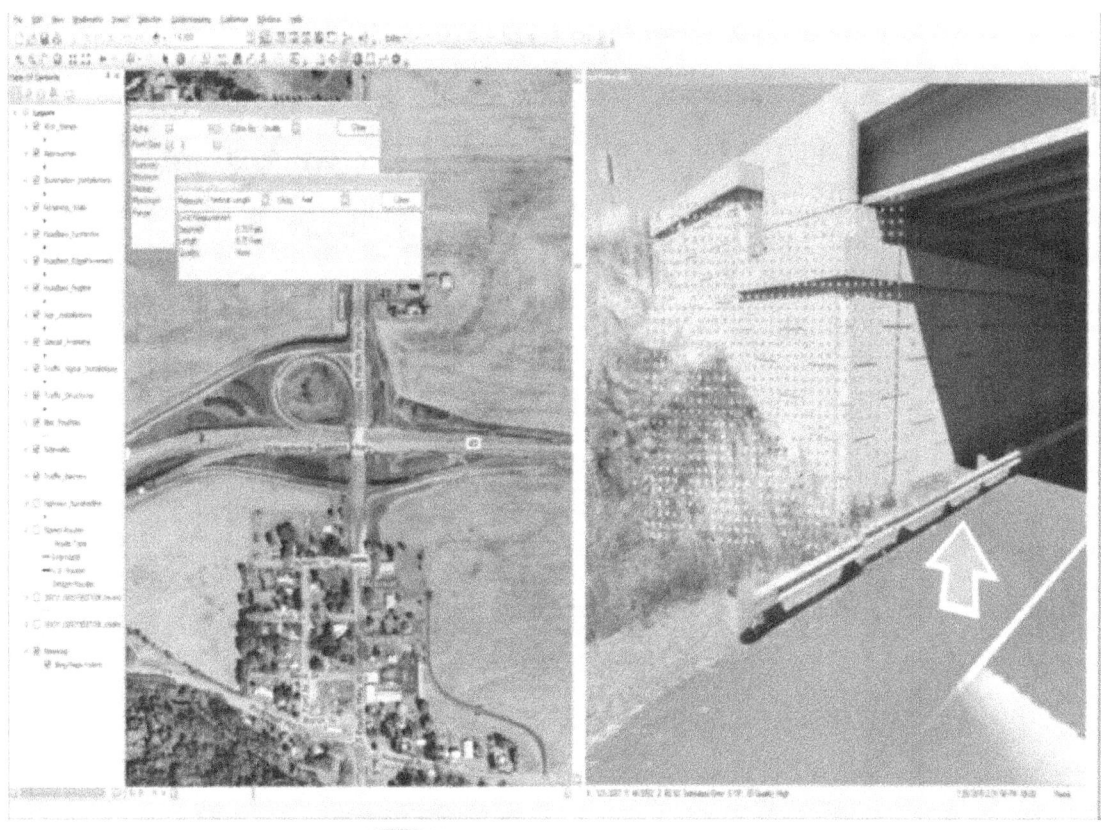
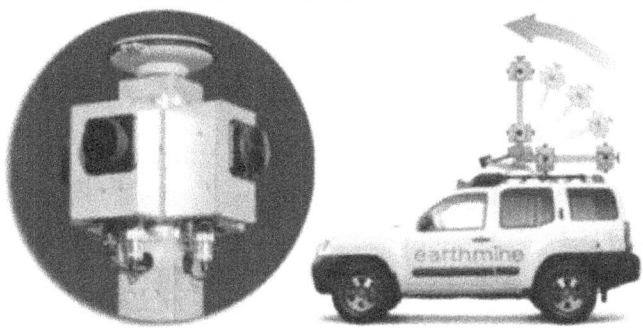

Figure 9: Earthmine for ArcGIS (top) is a mobile interactive mapping system that allows Oregon DOT engineers to "collect" asset data at their desks using video of the roadway collected from instrumented vehicles (bottom). Source: ODOT.

As part of **WSDOT's** RFIP, there is a custom mobile data collection system, the RFIP Data Editor. This ArcMap extension allows staff to collect data from their personal computers using images from WSDOT's State Route Web (SRview) database. The RFIP Data Editor allows users to view previously-collected video (see photo on the far left of Figure 11), then using the Editor tool, click on the start of the feature and record data for the feature. The result is a GIS layer showing the new feature (see Figure 11 bottom graphic). The state has future plans to add other data sources such as ortho photography and as-built drawings and to provide functionality to update existing data.

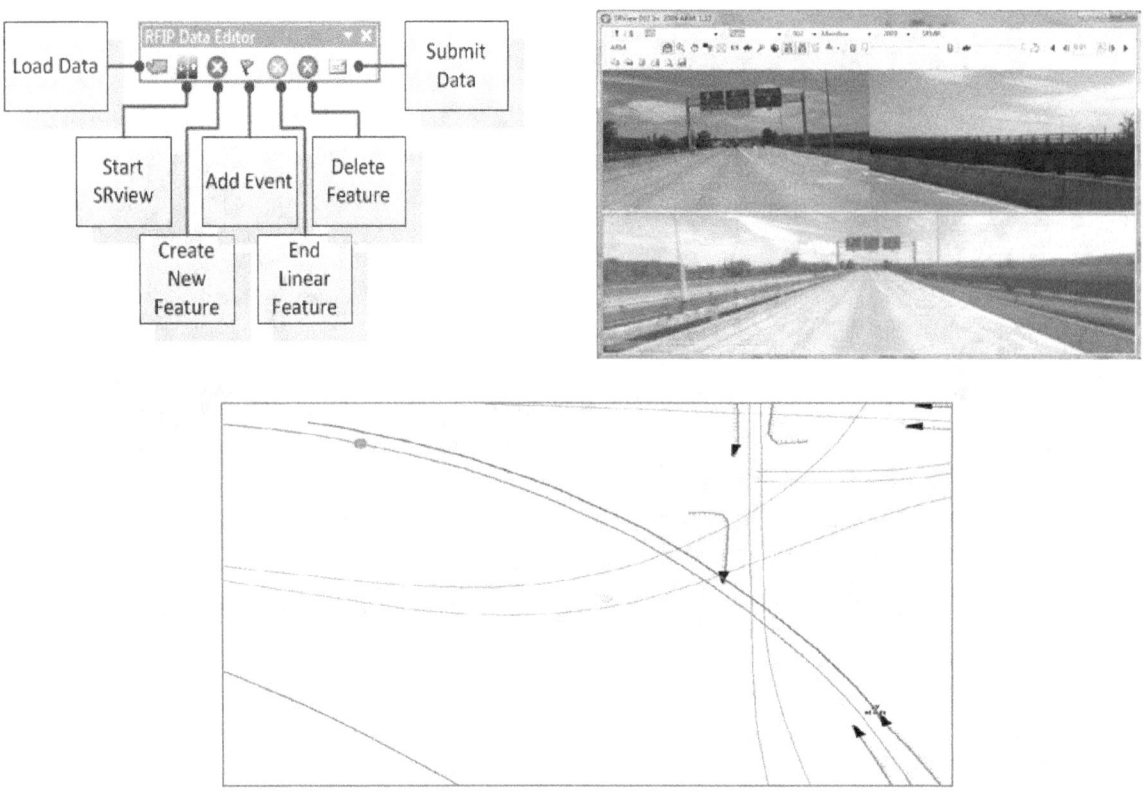

Figure 10: Screen shots from WSDOT's Road Feature and Inventory Program, a system used to collect data from recorded video. Source: WSDOT.

2.10.2 Data Storage

Another technical aspect of GIS and TAM integration is data storage. Data collected must also be stored for future use. COTS GIS and TAM software packages have custom databases designed to store data related to their respective purposes. Historically, data was stored in paper files or static electronic records that limited its transferability and required a human to physically enter data into electronic formats before doing any computerized analysis. Some data is still collected in paper form at many agencies and entered into databases as needed. However, with electronic data collection methods becoming more popular and affordable, transportation agencies are able to collect and store more extensive databases of data directly readable (or readable with minor conversions) by GIS and TAM systems.

More data is helpful for TAM and GIS, except that data storage comes at a cost. Servers are required to hold data; more data means more servers for the agency to purchase, power, and maintain. Infrastructure is needed to make stored data available to the people who need to use it. Ideally, agencies would maintain data in a coordinated and organized way, by eliminating redundant data, optimizing storage techniques, and improving database functions to minimize the need for network infrastructure. Most interviewed agencies acknowledged that data storage can be challenging.

2.10.3 Data Analysis

A third aspect of GIS and TAM integration is data analysis. The analyses that can be performed with digital data are some of the most attractive features of today's TAM and GIS technologies, and software developers in both fields have incorporated these capabilities into many commercially available software packages.

TAM analytical tools are primarily used to project future asset conditions and maintenance needs. It is also able to forecast budget information and develop asset condition scenarios based on changes in funding. Output data from TAM analysis tools can be output as reports or exported to other systems, such as GIS for viewing.

GIS analysis is used for location-based data queries, statistical analysis of datasets for a geographic area, and optimizing solutions. The COTS modules available to assist with these analyses typically come with user interfaces and embedded algorithms that allow agencies to alter the parameters. The outputs are often visual in nature (e.g., shading of regions that meet search criteria).

2.10.4 Information Dissemination

The final output of an integrated GIS and TAM program is information that can be viewed by internal staff and shared with others. It is common for TAM data to be communicated through charts, graphs, reports, and spreadsheets that can be published in hard copy or as static digital files internally and on the internet. Agencies can also develop query tools for live searching of TAM databases. GIS technology allows agencies to develop static maps and show a variety of dynamic data layers. GIS viewers can also make GIS data available to agency employees through intranets or to a wider audience over the internet.

2.10.5 Examples of System Architectures

The technical integration of GIS and TAM is achieved by combining the four aspects of technical integration – data collection, storage, analysis and dissemination – in different ways, depending on the needs of an organization. Several examples of GIS/TAM integration are illustrated in Figure 12 through Figure 14, each with unique features. MDOT's GIS system (Figure 12) represents a basic approach to using custom GIS software to display static data from TAM applications. WSDOT's Maintenance Operations (Figure 13) is a dynamic implementation that uses GIS to show real-time asset data collected from their snow-plows. Finally, St. Johns County (Figure 14) integrates asset management and maintenance management information with GIS and uses the analytical capabilities available through GIS in order to do its maintenance planning. *(Note: The figures are provided for informational purposes only and are not necessarily representative of the entire GIS or TAM programs at the agencies. See case studies in Appendix A for more information)*

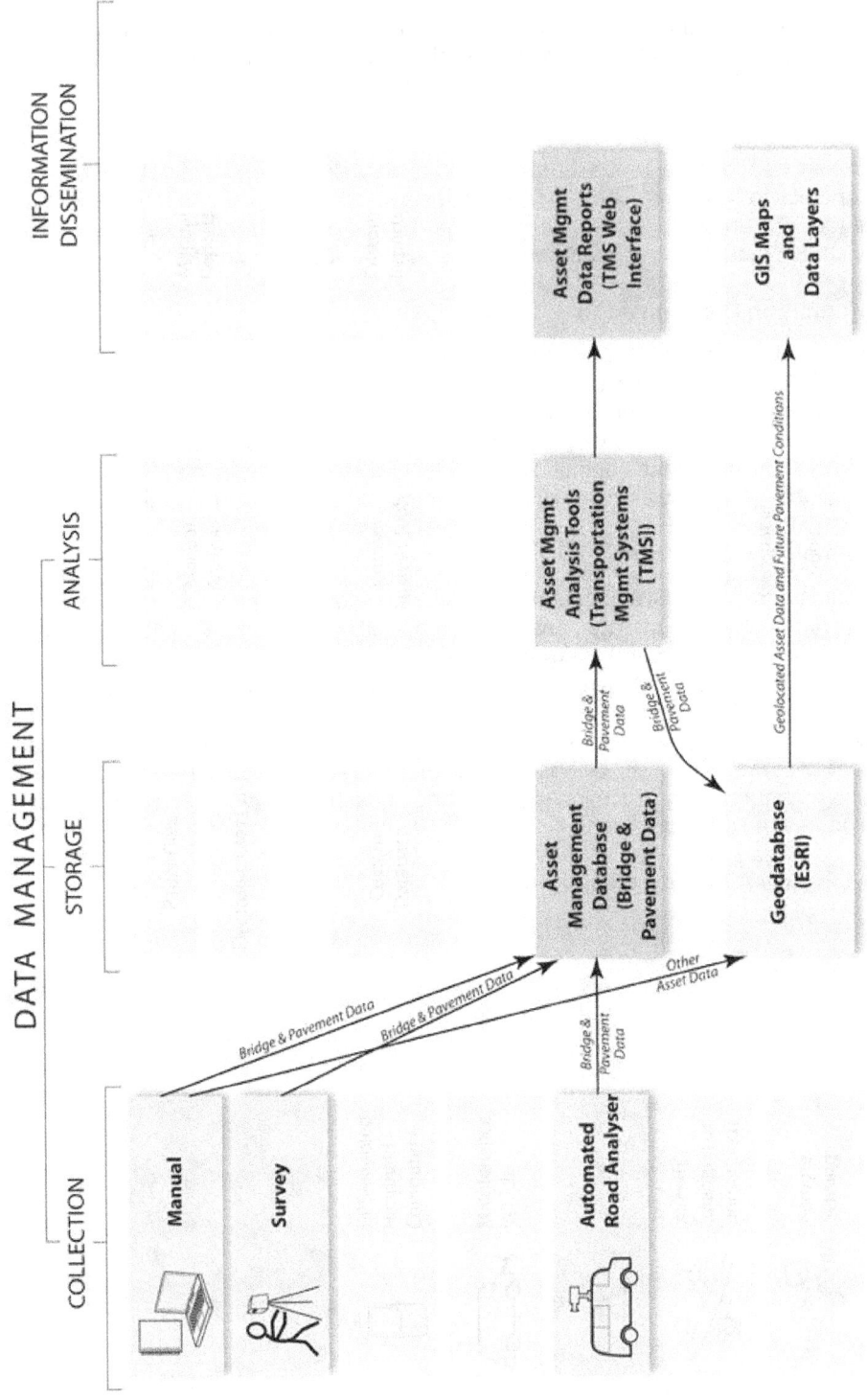

Figure 11: Michigan DOT uses GIS software to display static data from TAM applications.

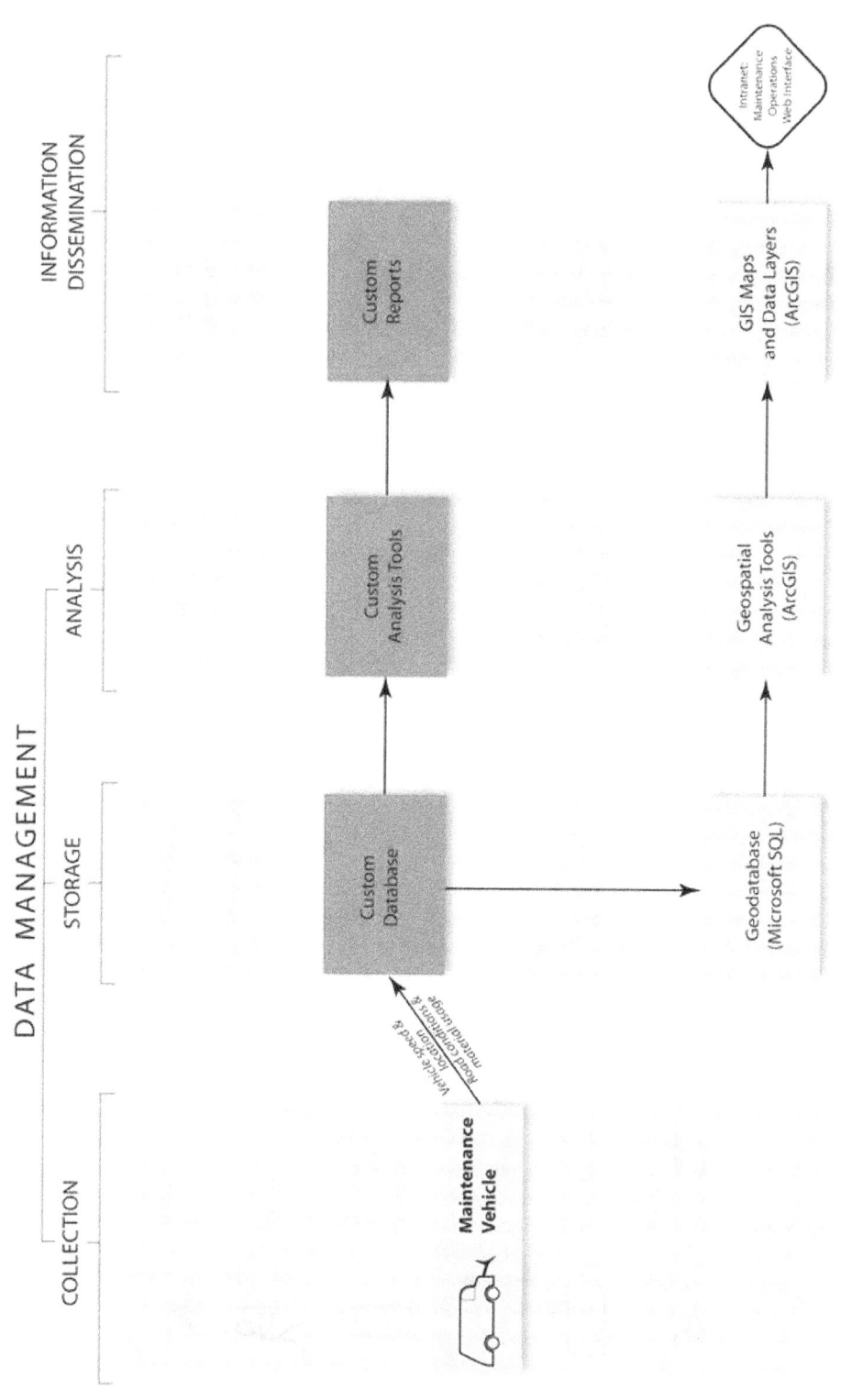

Figure 12: Washington DOT's system displays real-time asset data from maintenance vehicles on GIS maps.

St. Johns County Florida: GIS-based Maintenance Mangement

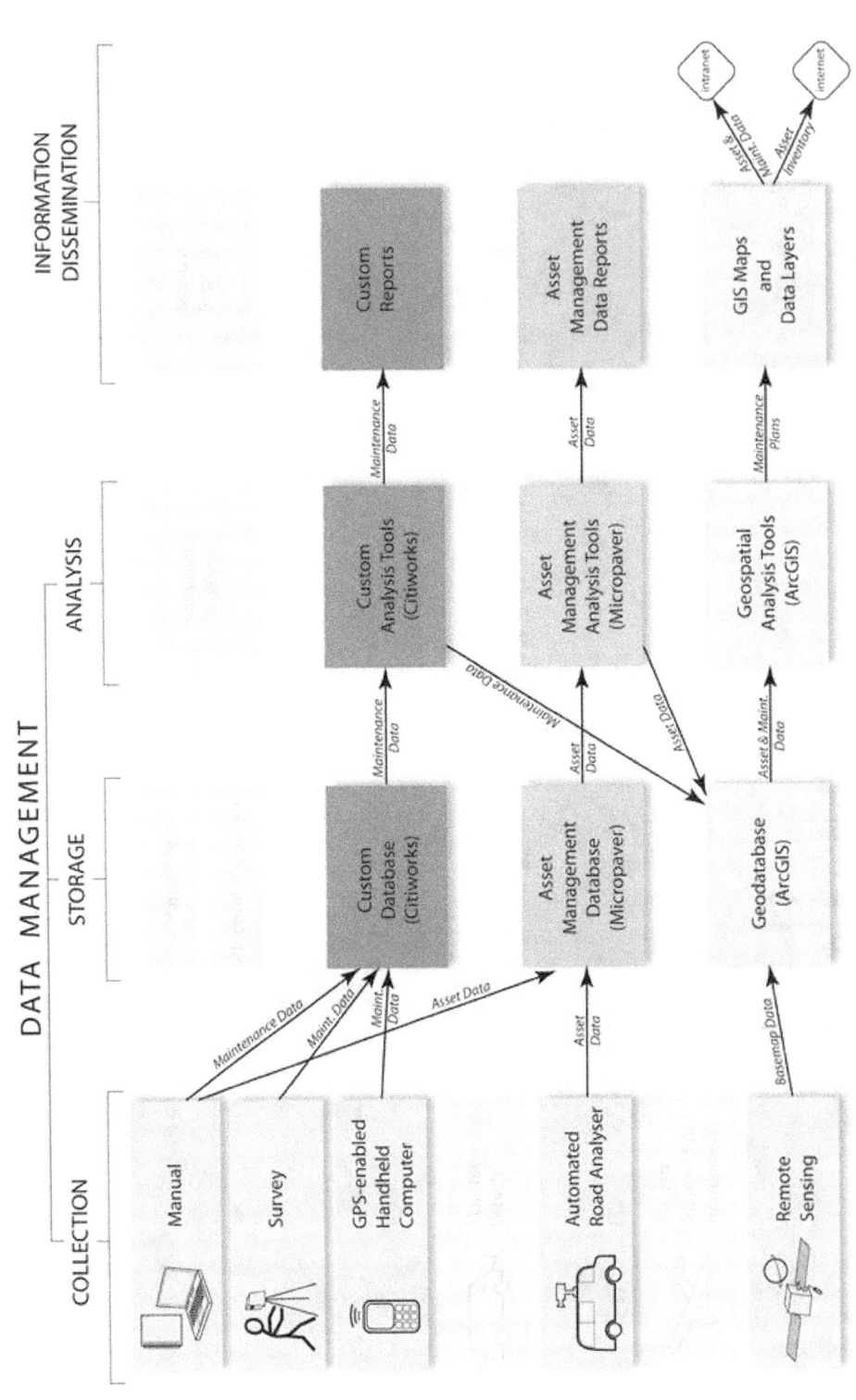

Figure 13: St. Johns County integrates asset management, maintenance management, and GIS into one system.

3 CHALLENGES

While every agency interviewed for this study has made strides toward integrating GIS with TAM, all of them acknowledged the challenges of starting and implementing these programs. The following section discusses these difficulties and the efforts agencies have made to overcome them.

3.1 STOVEPIPE ORGANIZATIONS

In most state DOTs interviewed for this study, GIS and TAM groups are functionally separated. The two functions may have different physical locations, separate budgets, separate leaders, unique functional roles within the business of the organization, and distinct work cultures. The separation can make coordination and sharing very difficult, especially at a budgetary level. It can also result in the duplication of effort if both groups are inadvertently collecting the same data. WSDOT discussed this concern: Two groups sometimes collect similar data for two different purposes; in the future, they aim to coordinate better so that both groups can get what they need from less data collection.

Agencies have overcome these challenges by creating ad hoc relationships across groups, by encouraging leadership of both groups to understand the complementary nature of their work, by hiring staff with more than one skill set (engineers with GIS capability, for example), by creating organization-wide data standards and data collection plans, and by making data from both fields available on cloud networks or intranets so that it can be used by groups across the organization. At Michigan DOT, for example, the TAM group has GIS specialists that help bridge the gap to other parts of the organization that use GIS.

3.2 DIFFICULTY GARNERING OR SUSTAINING LEADERSHIP SUPPORT

The support of transportation agency leadership is important for integrating GIS and TAM programs. Leaders advocate for funding, approve projects, enforce policy changes, and use the results of the program. Even so, it is not always easy for champions of programs to garner support from the executive leadership to invest in the software and staff training to support GIS, TAM, and integration between the two. Most agencies do not have a tangible way of measuring benefits to justify the expense of created integrated tools. Executives may be unfamiliar with the technology and apprehensive about implementing unproven applications. Their hesitation may be exacerbated by forced staff reductions and shrinking budgets that affect many states and municipalities. Travel restrictions and small training budgets also limit the ability of managers to commit to integrated GIS and TAM programs. Ohio DOT staff remarked that it is not always easy to get buy-in from leadership to invest in the technology, but executives and decision-makers have enjoyed the results, seeing technical data presented on easy-to-read maps.

> **Management Support for GIS in St. John's County, Florida**
>
> When St. Johns County, Florida Public Works Department was considering contracting out its mowing operations, management asked whether it would be more cost-effective than in-house work.
>
> Staff collected bids from outside contractors and pulled data about current in-house costs from the CMMS system. They compared the two and found in-house costs were lower in part due to the use of GIS to logically plan preventative work.
>
> The result was support from management for GIS-based maintenance management and for in-house maintenance as well.

3.3 DATA COLLECTION COSTS

Data collection is a critical part of TAM and GIS programs, but it is also a major obstacle to the successful implementation of GIS for transportation TAM. The time and cost of collecting inventory information for the volume of assets that are under the jurisdiction of most agencies can be overwhelming. Most agencies struggle with the question of how much data to collect except for assets that have state or federally-regulated data inspection requirements, such as Interstate highways.

3.4 COMPLEX SYSTEM ARCHITECTURES

Many of the GIS programs and TAM programs have complex system architectures independent of one another. Therefore, integration can be complicated. Some agencies are working through this by replacing old legacy systems with systems that are more integrated from the beginning (COTS TAM software that can export to GIS, for example). Even integrated systems require software and version upgrades, which can add significant cost to a GIS program. Often, one software package upgrade requires updates to linked software in order to maintain compatibility. Some agencies have simplified their architectures developing in-house software solutions that can integrate many types of systems and data. Others have selected a single vendor for all systems, minimizing the number of software packages required.

3.5 OWNERSHIP AMBIGUITIES

Data sharing between GIS and TAM raises questions of data ownership. Who is responsible for collecting data and analyzing it? Who is responsible for updating it? Who has the budget to purchase and maintain the systems? Sometimes agencies would like to coordinate GIS and TAM efforts, but they are unable to decide who will own and pay for the system, so the program is not established.

3.6 LACK OF DATA STANDARDS

GIS and TAM programs are affected by several data standards issues.

One issue is that different types of data are collected for different asset classes and at different regions at the state DOTs. The frequency of data collection may be different (e.g., every year, every 2 years, as-needed), the level of granularity may be different (e.g., every bridge joint vs. the entire bridge, or every mile of pavement versus every 0.25 mile of pavement), and the referencing system may be different (e.g. latitude and longitude versus milepoint along a road). Over time, agencies have found that disparate systems for data collection can become prevalent and ingrained, making it very difficult to combine the data into a centralized, state-wide system. Some agencies deal with these issues by creating data management plans and standards. Michigan DOT is working on such a plan in 2011. Similarly, CDOT is using an innovative approach to setting data collection standards by working with its procurement department to determine which equipment and software regions should purchase. This will reduce the varying number of inputs that need to be standardized into the database and reduce the overall cost of data integration issues occurring from many disparate systems. Another way that agencies have overcome this issue is to create a single Linear Referencing System (LRS) for all TAM and business data. A consistent LRS ensures that assets are consistently located along transportation routes.

The second standards issue is with GIS software data. From a GIS perspective, one private company with private data standards leads most of the market for GIS products. Other companies and organizations offer alternative data standards, but they are less common and less widely known. Many agencies use the dominating standard or convert their data to a compatible format in order to share it with other agencies that use the well-known standard. Interest groups are working to increase the adoption of open data standards, making it easier for companies to enter into the market. Diversification is expected to give agencies more choices about which software packages to use and allow them to purchase modules from different software providers that will work with the same systems.

A final data standards issue is related to conversion between GIS and asset management system data. Many COTS TAM software providers are designing their systems (or offering modules) to export TAM data to a GIS-friendly data. As it stands, many legacy systems did not have this capability, which means that data must be converted once (or sometimes more than once if it is converted at the point of entry into data storage systems) in order to use it for GIS purposes. This can be time-consuming, especially for data that must be updated regularly.

Michigan DOT's Data Standardization Rules Enable GIS Based Sharing of Data Across DOT and Counties

The Transportation Asset Management Council (TAMC) was established to expand the practice of asset management statewide to enhance the productivity of investing in Michigan's roads and bridges. Act 499 of the Public Acts 0f 2002 defined asset management, established the TAMC, and identified roles and responsibilities of the TAMC and member agencies. TAMC is a legislated body of representatives from transportation agencies that coordinates the collection of condition data on Federal-aid-eligible roads and bridges, collection of asset investment data and the reporting of the collected data to the Legislature and State Transportation Commission.

Part of the TAMC's mission is to gather physical inventory and condition data for all roads and bridges in Michigan. Their online portal, the Investment Reporting Tool, is designed to accept past, current, and future investment data from any road-owning agency in Michigan. Data collected is used to report on the current condition of the roads and bridges and also to predict the future condition of the road network of the state. Data must be reported each year even if there were no treatments performed. Also, all data submissions must be reported to the TAMC using specific data standards. The data is then stored in a central data agency database. The database is also the source for all applications storing, editing, or accessing data for TAMC purposes.

One central location for data allows several agencies in Michigan to work cooperatively rather than competitively. Information can be produced for distribution to the public from one central location. TAMC has an updated website called the Information Portal that is intended to be more user-friendly for the general public and the 600+ Michigan road agencies completing annual reporting requirements via the Investment Reporting Tool. The site also provides interactive maps enabling the public to view pavement and bridge conditions. There is also a dashboard feature that shows actual bridge and pavement condition data for the years of 2004 – 2010 and forecasted for 2011 – 2015. All conditions are listed as good, fair, or poor.

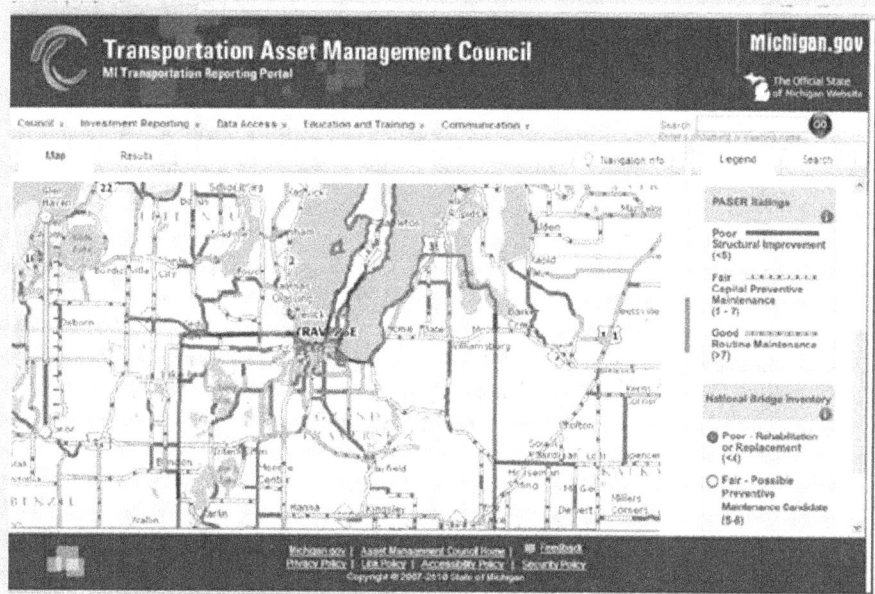

Image: TAMC website.

> **Colorado DOT's Policy to Develop a Single LRS**
>
> In 2005, CDOT's Division of Transportation Development (DTD) initiated a policy to develop a single LRS for the entire Department so that all location-referenced data collected on several corridors would be consistent and easily integrated with other corridor data. The single LRS was able to unify many different LRSs that had been developed over the years between three business units and six regions into one linear reference framework for all CDOT highways. CDOT has completely retracked its highway assets for inventory purposes over the last two years using the single LRS. The unified LRS is now being used as the base layer for enterprise data collection.
>
> The LRS policy was particularly important when the agency implemented a new SAP-based GIS and TAM system, because the unified LRS effort produced matching data segmentation for all highways and data sets. The single LMS has enhanced the geometric accuracy of the data. Each route was compared using beginning and end reference points and length in the three different databases. It was then matched against historical highway logs and evaluated using current data such as aerial imagery, digital video logs from the pavement management system, and as-built drawings provided by the regions, among others. Using a single inventory and common data source for all business units, SAP will enable consistent financial planning and management.

3.7 STAFF SIZE AND TIME CONSTRAINTS

Most agencies have limited staff dedicated to GIS and asset management, and many of the existing employees are very busy as it is. In the current economic climate, staff training and travel budgets are also limited constrained. As a result, agency staff members are limited in their abilities to expand GIS programs and intentionally integrate them with TAM programs. Michigan DOT, even with over 50 people in its Asset Management Department, struggle with finding people to help its regions with GIS. That said, some individuals have found ways to stay apprised of the industry by participating in web-based collaborative groups, such as the GIS for Strategic Asset Management (GSAM) project (see Section 5.2, "Collaboration Among Peers", for more information about GSAM).

3.8 DETERMINING HOW MUCH DATA TO SHARE

GIS can be a powerful tool for involving the public in transportation decisions. However, transportation agencies may not always be comfortable with giving the public large sets of data, especially those that could lead to security threats. WSDOT mentioned that it is working through this issue as it considers making asset data publicly available via a GIS application in the future.

4 TRENDS AND NEW TECHNOLOGIES

TAM and GIS systems continue to evolve. It is a challenge for agencies to stay apprised of developments as discussed in the previous section. Some new technologies and industry trends relevant to GIS and TAM are described in this section. Please note that although many proprietary products are described in the section below, FHWA does not endorse any of these products particularly. These technologies were individually selected by each agency or DOT.

4.1 IMPROVED MOBILE DATA COLLECTION

Mobile devices are becoming more powerful computers and data collection instruments. Currently, they are most commonly used to *record* data collected through other means (e.g., manually, by an inspector), as discussed in Section 2.10.1. In the future, mobile devices themselves will likely be measurement instruments that collect and enter data directly into databases. Mobile devices can also be configured to automatically geolocate the data collected based on the GPS readings on the phone, reducing the need for data entry and ensuring that collected data is properly associated with its true location. These capabilities make mobile devices likely candidates for use in GIS and TAM.

There are already many GIS applications for mobile devices, both for-profit/closed-source software that are purchased as is (e.g., ArcGIS for Windows Mobile, Virtual Earth Mobile) and open source solutions that allow users to access and edit the source code (including Bluemapia, gvSIG Mobile, JVNMobile GIS, and others).[6] Several asset management software providers also provide mobile device compatibility for data entry (e.g., AgileAssets, Cartegraph).

There are also GIS-based mobile device applications that allow members of the public to report asset information. The City of Boston, Massachusetts has two such applications. The first is called "Citizens Connect" and it allows the public to use their mobile phones to report issues (such as graffiti or potholes) to the city. The second is a smartphone application called the Boston Urban Mechanic Profiler (StreetBump) that is designed to use the sensors installed in commercially available smartphones to record gross pavement roughness when a user rides over a section of road. The "bumpiness" data is collected and stamped with the user's speed and location and made available to the public (see Figure 16). The application is currently limited by the accuracy of the sensors and is under refinement to convert the data collected into useful information,[7] but it is an example of geolocated TAM data collected by the public using a mobile device.

Involving the Public with Data Collection

The City of Boston introduced "Citizens Connect," a cutting edge, full-featured tool for enabling Boston's residents to improve its neighborhoods by reporting issues such as potholes, broken street lights, and damaged sidewalk patches. Users can download the mobile application (App) to their phones. The App enables residents to send both pictures and text with a feature that tags that information with location-based GPS. After filing a complaint, users will receive a tracking number so that they can easily follow up if the problem persists. After a request is submitted, a work order is immediately assigned. In June, the App was updated to allow constituents to report a damaged or missing sign.

Image: Citizens Connect Website

---

[6] OSGeo,"Mobile Solutions." Website. Last updated 21 Jul 2011. http://w.ki.osgeo.org/wiki/Mobile_Solutions#Open_Source
[7] Mayor's Office of New Urban Mechanics, The, "Street Bump – Features." Website. www.newurbanmechanics.org.

As technology continues to advance, mobile devices will offer the opportunity to reduce the cost and improve the flexibility of data collection for TAM and GIS. Meanwhile, there are several limiting factors and challenges that need to be addressed, including hardware and sensor accuracy, data standards, discrepancies across smartphone operating systems, network connection issues, and GPS accuracy limitations.

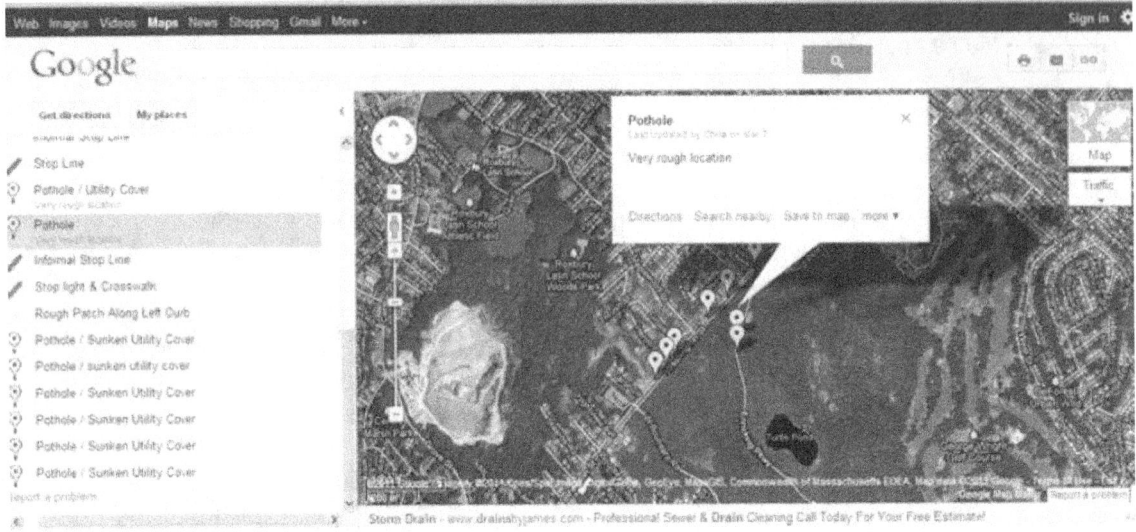

Figure 14: The City of Boston mobile device application, StreetBump, collects geotagged rough road profile data (top) that can be used to show potholes and other roadway irregularities on a map (bottom). Source: StreetBump website.

4.2 Automatic Geotagging

Traditional asset management data collection devices are being designed with in-built GPS technology to automatically geotag the information they collect. For example, many digital cameras now have a geotagging option that allows photographs to be automatically geolocated when they are taken. When the photographs are uploaded to a computer they can then be viewed, categorized, and associated with other information based on the location at which they were taken. Ohio DOT District 2 geotags photographs taken with its digital camera (see Figure 16).

4.3 Cloud Computing

Cloud computing is becoming an increasingly popular way to remotely store and share software and data among many users. Much like large-scale servers, "cloud" networks are accessible anywhere in the world (or anywhere on a transportation agency's property) through the internet. Agencies typically pay cloud computing companies service fees to host their data and then agency staff can log on to access the data from any internet-enabled computer. Some software companies (including those that produce COTS asset management software) are also beginning to offer their own cloud software services. In this context, agencies typically pay for a set number of users, who can then logon from any internet-enabled computer.

The advantages of storing data or using software in the cloud include:

- Potential cost savings of paying another entity to operate the service rather than maintaining server infrastructure in-house;
- Reduction of redundant data through centralized version and content controls;
- Ease of sharing data among many different computers and users on different networks; and,
- Increased use of data standards to facilitate sharing among different organizations.

Figure 15: Ohio DOT District 2 geotagged photograph displayed on a map. Source: GSAM.

The main disadvantage to cloud computing is that private companies are responsible for data storage and services, which gives public agencies less control over the security and safety of their data. Agencies may also be more committed to a vendor that stores their data in the clouds. Agencies can minimize risk by ensuring that they have proper contractual protections in place when working with cloud providers and maintaining backups.

4.4 INCREASED USE OF AUTOMATED ANALYSIS

Another future trend will likely be the increased use of GIS analysis tools that utilize geospatial data to correlate asset data with geographic data (e.g., topography) and/or other data sets (e.g., socio-economic data). Automated GIS analysis tools can allow agencies to develop better designs and more efficient operating and maintenance plans, assess asset performance against safety measures, undertake disaster planning, and otherwise enhance traditional TAM programs by automatically analyzing asset data against other data sets. As mentioned in early sections, this capability exists in GIS software tools but not many agencies use it on a regular basis.

4.5 3-D Visualization

Three-dimensional (3-D) visualization is another trend that will influence the fields of TAM and GIS, especially as it relates to monitoring large structural assets such as bridges, buildings, and signs. All GIS software packages are designed to project 3-D landscapes, but only recently have commercial GIS software developers begun to incorporate 3-D graphic capabilities for displaying assets in a virtual landscape. The 3-D visualization capability can be used to show even more information to engineers and decision-makers in ways that are easy to interpret. For example, rather than showing 2-D colored lines on a map to represent pavement condition, a 3-D visualization might show the texture of the pavement in question and the extent of damage on it; similarly, 3-D visualization could help project planners and the public see and "feel" the difference between an at-grade roadway cutting through a wetland versus a more expensive but less intrusive alternative that circumvents the sensitive land.

4.6 Visualizing Future Conditions

Most agencies interviewed use GIS to view current conditions. However, some agencies use GIS to visualize the future. MDOT is able to show future conditions of its pavement segments, based on projections provided by its pavement management tools and georeferenced in its GIS department. In the future, more agencies will likely map future asset conditions to help decision-makers and the public understand the impacts of present decisions on the future.

4.7 Dynamic Segmentation (DynSeg)

Dynamic segmentation, often shortened to DynSeg, is a GIS feature that allows tabular data to be associated with static geographic features within GIS based on a common linear referencing scheme rather than be mapped independently. The linear-referenced, tabular data are called events. These events are geolocated based on their relationship with the linear data. For instance, a 25-mile road can be geolocated based on its global coordinates (latitude and longitude). Using DynSeg, it can also be displayed as one linear feature with a starting point and ending point. Then event data at a specific location, such as an accident at mile 16, can be displayed as a point on the line and does not require its own global coordinates. An event can also cover distance. Using the same 25-mile road as an example, a segment of highway re-paved between mile 3 and mile 6 can also be stored as an attribute of the road using DynSeg.

The primary advantage of DynSeg is that it allows multiple types of linear data related to single features to be stored in GIS without duplicating base data or dividing the feature into many unwieldy segments. While DynSeg has been around for many years, it has more recently become a standard tool

---

3-D Visualization at Ohio DOT

Transportation networks are often represented by a two-dimensional plane and are limited to basic attributes of topological distance and shape. LiDAR datasets are used for three-dimensional (3-D) transportation network information to be created by the GIS application. The Ohio Statewide Imagery Program (OSIP) is a partnership between state, local, and Federal government agencies to develop high resolution imagery and elevation data for the State of Ohio to benefit GIS users at all levels of government.

With OSIP, a new statewide collection of color ortho-imagery will be created over the next four years (2011-2014). With the implementation of remote sensing technologies, the new ortho-imagery provides a strong base for which the extraction of ground features (i.e., woodlands, tillable, impervious surfaces, etc.) can be performed. For example, ortho-imagery combined with GIS can provide a 3-D visualization of an elevation profile where a new highway is proposed. This method saves time and cost by limiting field data collection, but it also improves accuracy. For the roadway scenario, it can also be used for predicting flood areas and landslide modeling based on the placement of the highway.

Image: Ohio DOT

in most GIS packages. The capability has numerous applications to transportation and asset management because most transportation infrastructure is linear (e.g., roads, rail, pipelines) and can have many different types of data associated with it (e.g., a road can have multiple bus routes running over it, accident data for some points, and height limitations at underpasses). Most transportation infrastructure is also directional. DynSeg allows a road to be referenced in both directions (i.e. from mile 3 to 6 and 6 to 3) and thus associate tabular data with a specific direction. As practitioners gain the knowledge and expertise to execute proper DynSeg within GIS, its advantages will be seen throughout any TAM program that involves linear features.

4.8 OPEN DATA

The trend toward open data (that is, the willingness of agencies to share raw transportation data with the general public), will also likely affect GIS and TAM. Already, many agencies provide their GIS base maps and asset inventory layers on the internet. Many agencies also make operating data (e.g., real-time bus locations and traffic camera views) available. In the future, agencies may provide more data, such as condition or performance data, for others to view and use. Some agencies, such as WSDOT, are already discussing the benefits and risks of opening their GIS and asset management data to the public. They are working to answer some key questions: What types of data would be useful to the public? Are there any types of data that should not be released due to their safety or security-sensitive nature? What is the best way to release the data?

5 ONGOING RESEARCH

There is also ongoing research into GIS and TAM. GIS is a diverse field, requiring expertise in a wide range of areas from cartography, systems administration, database management, programming, and spatial analysis as well as expertise in how GIS can be related to TAM. Even the well-versed GIS professional may come across many difficult tasks that require guidance from others in the field. A strong support community has been building itself into a readily accessible network, both via the Internet and in person through focused user groups and forums, to assist with day-to-day GIS and transportation management activities. Outlined below are resources that GIS professionals are utilizing to help guide them through their technical questions and connect with other GIS professionals.

5.1 ACADEMIC RESEARCH

Researchers are exploring new applications, philosophies, and procedures for GIS and TAM as well as new hardware and software developments. Many researchers are currently focused on integrated decision support – taking outputs from TAM inventory, condition assessment, and deterioration modeling and combining them with other sources of data in a GIS environment to enhance the assessment of infrastructure performance and improve the life-cycle decision-making.

Academic research done at the North Carolina State University developed a system for estimating the physical condition of pavement markings based on collected data and predictive algorithms. Managers can use either the measured data or predicted data the system generates to decide on the best possible marking to use with the given pavement type and environmental conditions. Ultimately, managers can develop cost-effective strategies for pavement marking asset management and view using GIS.

5.2 COLLABORATION AMONG PEERS

GIS for Strategic Asset Management (GSAM) Group

GSAM is an informal group focusing on issues that GIS professionals increasingly face with the collection and management of large asset inventories. The group structure is based on collaborative discussion and idea-sharing surrounding the use of GIS for TAM. The forum is open to government and educational agencies to engage in conversation about any TAM-related topics, best practices, and lessons learned. By coordinating a group of interested agencies, this provides the possibility to share knowledge and reduce duplication of efforts when developing asset management practices integrated with GIS.

To date, the group has focused on discussions such as common solutions to spatial asset data collection, exchanging ideas on how other agencies are tracking assets in order to evaluate the various cost effective methods best suited for their needs, and sharing of database designs and code examples for asset management. The members share a common goal in the importance of having conversations on these topics so that GIS professionals can do their jobs more effectively and efficiently in order to better manage data, design work flows, train field crews, support hardware and software, and analyze data.

The group is facilitated by staff from the Ohio and Iowa DOTs. Members include representatives from universities and several sectors of government including cities, counties, Federal, regional, and state agencies. The group holds quarterly meetings and maintains a site for collaboration and data sharing. Each meeting contains presentations from group members, third-party industry vendors, and applications of new technologies and innovations. This format allows for members to share their experience of what is being done at their organization and to gather feedback through discussion. This also keeps members apprised of the latest technology and industry trends.

AASHTO GIS-T Symposium

Since 1987, AASHTO has sponsored the annual GIS for Transportation Symposium (GIS-T). This provides an opportunity for both public and private industry interested in the use of GIS for transportation purposes to collaborate and share experiences, view the latest software and technology trends, and learn more about the field. The Symposium also provides an avenue for participants to network with peers to discuss emerging issues and concerns.

GIS-T offers a variety of keynote speakers, discussion forums, workshops, presentations, and a technology hall where exhibitors showcase their services. Organizations and individuals with information related to GIS in transportation are encouraged to share their experience by presenting at the Symposium. Each year, a summary report is produced that identifies key emerging issues and discusses how their impact might affect the GIS-T community. Input is gathered by means of a pre-symposium survey, session papers, panel discussions, and the Symposium wrap-up session. In 2011, the following key issues were identified: data sharing and coordination, mobile computing, cloud computing, architecture management, data visualization for the public, and using GIS for safety analysis.

5.3 AGENCY STAFF INVOLVED WITH RESEARCH AND DEVELOPMENT

Several interviewed agencies are also involved in ongoing TAM and GIS research. For example, Ohio DOT is developing in-house software to catalogue assets and implementing the use of aircraft surveillance for data collection (see page 17). WSDOT is expanding its Maintenance Operations system to include more functionality and data collected from other maintenance vehicles like mowers, line painters and vacuum trucks (see Appendix A, page A-6). The lessons learned and technologies developed through these efforts also contribute to the field.

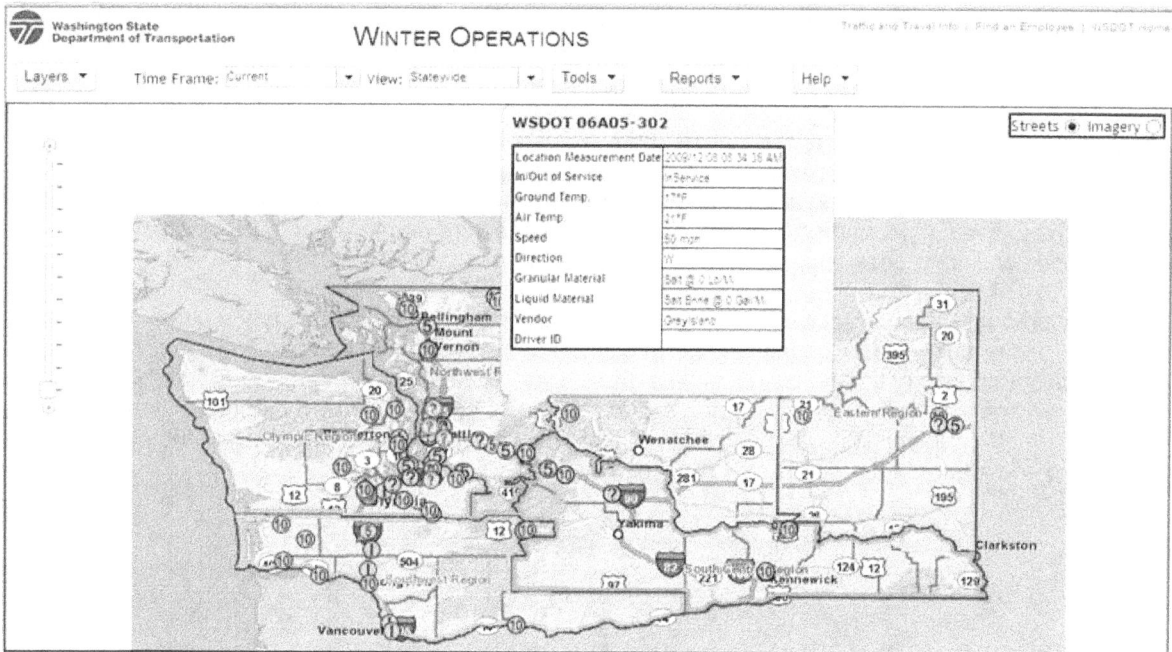

Figure 16: WSDOT's Maintenance Operations System (formerly Winter Operations) shows live information collected from plow trucks. Staff is working on upgrading the system and collecting data from other types of maintenance equipment (such as mowers, line painters, and vacuum trucks) Source: WSDOT

6 CONCLUSIONS

GIS enhances the field of TAM. Maps help leaders see the extent of problems, understand the geographic impact of their decisions, and ultimately make more informed decisions. Maps can also help the public see and understand the far-reaching importance of the transportation assets they use every day. GIS enables transportation agencies to show information about their assets on maps that both technical and non-technical audiences can understand. It provides analysis tools that agencies can use to consider geographic features in the maintenance and design of their infrastructure. It also allows agencies to compare asset data with socio-economic, environmental, financial, and other types of data to identify relationships that they may have not considered before and make better decisions.

State DOTs have come far in applying GIS to many areas of transportation. However, the true potential of GIS in transportation asset management has yet to be fully realized in all transportation agencies. The interviews conducted through this study suggest that there may be many reasons for this, including:

- Stovepipe organizations
- Lack of staff training and appropriate software modules (both tied to funding and leadership support)
- Limited data availability in both TAM and GIS
- Proprietary standards that make sharing difficult.

Even in mature GIS and TAM programs, good data, collaborative cultures, technical standards, and leadership support are necessary keys to success.

New technology offers solutions to existing challenges and provides advanced agencies with a foretaste of even more capability yet to come. New mobile devices will make data collection less expensive and easier to do. Ongoing TAM and GIS software improvements will make data manipulation and advanced mapping easier. Open source software and industry data standards may make the market for GIS and TAM software more versatile and affordable to small agencies. These small changes will make it easier for agencies to use GIS in their TAM programs but will also require practitioners to follow changes in the industry and continually grow their skills.

State transportation agencies aim to provide safe, reliable, and efficient services to the users of their systems. They are also held accountable for the decisions they make and the taxpayer funding that is used to pay for maintenance and construction of the system. Taken independently, TAM and GIS activities can help agencies to provide better service and to do so in an efficient manner. Taken together, GIS can help staff communicate the findings from their TAM programs in visual way that the public can understand and leaders can use to make better decisions.

APPENDIX A: CASE STUDIES

ST. JOHNS COUNTY, FLORIDA

Background

St. Johns County is located in northeastern Florida, stretching along the coast of the Atlantic Ocean, with a total area of 821 square miles and a population of 190,039 persons. St. Johns County's main economic base is tourism, and thousands each year visit the approximately 43 miles of beaches and the historic county seat of St. Augustine. Being part of the Greater Jacksonville Metropolitan area, St. Johns County has experienced explosive growth in the last 10 years as people working in Jacksonville have built new homes within the County. The ocean-front location also lends itself to the threat of hurricanes each year.

In an effort to keep ahead of growing demand of residents and visitors, the St. Johns County Public Works Department works to maintain, preserve, and protect the infrastructure resources of the County in the most efficient and effective manner possible. In 2005, the Public Works Department implemented stand-alone GIS datasets and databases primarily for roadways to begin tracking assets. The Department then conducted a full review of its business operations, including TAM practices and maintenance operations, to determine how its processes could be improved and transitioned into a performance-based system. The study found that 75 percent of maintenance being performed was responsive and only 25 percent was proactive. With such expanded population growth, the Department was looking to reverse this trend and to also create a transition of customer expectations.

The outcome of this review was the idea to develop enterprise TAM system (EAMS) based solely on GIS. The foundation for the EAMS would be a CMMS that would directly interface with a fully populated GIS database. The Department selected the CMMS as the first software to integrate with the GIS database because its primary function is to monitor the workflow of the Road and Bridge Division, which is the most labor intensive of the four public works divisions. This division maintains and improves the County roadways and drainage systems and is also responsible for all county mowing. Also, maintenance management was not new to the County - the new GIS-based CMMS replaced an existing non-GIS based CMMS. It was critical to select CMMS software that was built directly on a GIS geodatabase so that in the future, St. Johns County could continue to use the single, editable data source for all St. Johns County departments, not just public works.

Implementing a Maintenance Management System Integrated with GIS

In 2005, the Public Works Department began implementing the new CMMS. The Department's main goals were to create and track service requests and work orders, effectively locate the assets associated with the activity, and account for the costs by activity. Another goal was to use a single GIS system for financial management and TAM, thereby improving financial accountability and operational efficiency.

Before the new maintenance system could be implemented, asset data was collected from the entire county, including all transportation and drainage assets, which were then entered into one GIS database. This included signs, traffic signals, guardrails, culverts, ditches, bridges, pavement, and sidewalks – basically anything that the County is responsible for inspecting and maintaining. A consultant was hired to collect the asset data and populate the geodatabase using various collection methods. Since the CMMS is integrated with the GIS database, data can be entered through either the CMMS interface or directly into the GIS database. Costs of labor, materials, and equipment are entered into the CMMS system, which is then linked to a particular asset in the GIS database. Condition data can also be entered into the system. Currently, the database contains condition data for signs, culverts, pavement, and traffic signals.

Having maintenance software completely integrated with the GIS database means that there is no reliance on a third-party software to have the systems share information. Maintenance work orders easily attach to asset data with no import/export mechanism required. Also, all systems are fully integrated and connected within the network, including desktops, laptops, and field computers. As new assets are constructed, data is collected by hand by the Department and entered into the system. It is currently in the process of upgrading its system to deploy field units for engineers to use.

Using GIS to Improve Efficiency

Daily maintenance is now based on a proactive, methodical approach in which the County is divided into four work zones with each zone having a designated team that is responsible for all routine maintenance within that particular zone. Work is scheduled two weeks ahead and mapped out by area along a particular segment of roadway. GIS enables query of all assets that fall within the designated area. The County then plans its work as a "sweeping" motion where maintenance of all assets that fall within the mapped area is carried out instead of jumping from place to place or asset to asset. At each location, each zone team is responsible for all maintenance-related tasks, including culvert clean-out, washing/straightening signs, repairing reflectivity, etc. The sweeping motion saves on crew and equipment relocation costs. The County is also able to systematically track how often maintenance is completed for each asset. With this system in place, maintenance is now 80 percent proactive and 20 percent reactive. The reactive maintenance requests are handled by two rapid response teams that can respond to a particular concern or emergency request, such as a fallen tree or debris in the roadway, so that routine maintenance crews do not have to mobilize.

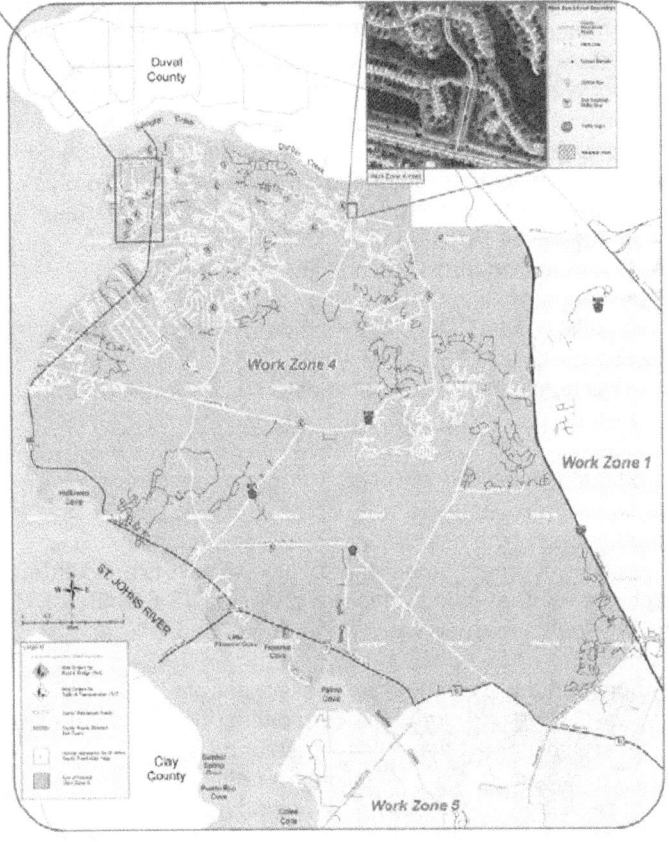

Figure A-1: Work Zone 4. Source: St. Johns County

St. Johns County has started using the GIS database and maintenance management system to reallocate work to different zones based on predicted work load. Since the area is growing so rapidly, new assets are being constructed on a regular basis; and some zones are growing at faster rates than others, particularly in the east and northwest sections of the County. The goal of the Department is to create balanced work zones. Each year, the assets of each zone are assessed. Then the zones are redrawn to balance the workload as well as the budget instead of being based on land area. This is an innovative way of using the system to use resources more efficiently.

St. Johns is also able to use GIS for performance budgeting. Information from the daily work orders from the CMMS application is reported through monthly work order reports, summarizing the total production of each work activity by geographic work zone. This provides a direct comparison of the actual work performed versus the work plan. The reports also include the amount of resources consumed, the amount of work accomplished, and the detailed cost for each work activity per geographic work zone. Using the ArcMap queries, it is simple to search and gather work activity information within any of the zones. Work activities are associated to a feature dataset, making it possible to narrow down specific work to an asset. For example, it is possible to search for all pothole repair activity for the fiscal year and display this information graphically. Using the same concept, the Public Works Department is prepared to track any hurricane debris removal activity and cleared roadways, allowing the department to inform County staff and the public in a timely and accurate fashion.

Challenges

Transitioning from tracking assets and generating maintenance work orders via paper required a proactive approach to training. First, the Department moved toward general computer usage, utilizing Microsoft Excel spreadsheets for tracking and scheduling work, recording time, and other basic tasks. As personnel became more comfortable with using spreadsheets for these tasks, they also realized the time savings associated with automating processes. Now, they have moved away from paper and Excel spreadsheets, and everyone is entering data either through the CMMS or GIS database.

The biggest challenge for the County today is keeping up with changing technology. There is significant cost associated with software upgrades for the CMMS and ArcGIS Server as well as the time that it takes workers to learn the latest software versions, which directly impacts workflow. Typically, one software package cannot be upgraded without upgrading the others as compatibility issues arise. While there has been some discussion of developing in-house software solutions that are more cost effective, the time that it would take to design, test, and roll-out a solution has been found to not be efficient in comparison to the cost savings. Once it has been decided that software will be upgraded, training is required for all users; and in recent years, funding for training has been drastically reduced.

Benefits

There are many benefits to implementing a fully integrated GIS database, including hurricane preparedness, assistance with yearly budgeting, and expediting the asset inspection process. At the start of the hurricane season, the County is able to overlay two maps – one that shows hurricane flood-prone regions and the other that shows where draining ditches and culverts have been recently cleared. This is used to ensure as little damage as possible to the region. Also, if the area continues to flood even after knowing that all of the waterways have been cleared, this may be an indication that an engineering redesign may be the only solution.

For inspection purposes, having all the assets inventoried and mapped saves significant time in the field. For example, now engineers can go into the field with a field laptop showing a clear map of where each culvert is located and not waste time searching for a culvert that may have been misidentified in a paper report. You can also print the previous condition report and compare the new conditions to be able to make a clear determination of what changes have been experienced since the last inspection.

To make maintenance decisions, management now has the ability to enable performance-based budgeting. The system can provide data for a particular type of road and the cost of maintenance on it. The Public Works Department can now budget for the entire year. It can also use this data during budget cuts – different scenarios can be generated to see what happens when a certain number of dollars are cut – what projects make the most sense, what projects can wait until the next budgetary cycle, etc.

Outcomes

Using GIS to link asset inventory, condition data, and financial information allows the County to answer questions such as "how much does it cost to do this type of maintenance?" or "how often are we doing this type of maintenance and can we do it less often and get the same result?" Combining this type of data allows for performance measurement and performance-based budgeting.

In the first full year of implementation (2006 – 2007), the system helped increase productivity by more than 13.3 percent, saving St. Johns County over $650,000. Prior to this program, the Department had no way of systematically estimating or predicting maintenance costs from year to year. After implementation, when the Department was considering contracting out the mowing operations, it was able to collect bids from outside contractors to compare to current internal costs. It was found that the internal operation costs were less costly than hiring an outside vendor.

St. Johns County is leading the way in TAM and maintenance operations for local governments. The system is being modeled by other local governments, and the Department has received many inquiries from surrounding agencies as well as agencies from around the United States asking questions about implementation, data models, and design schemes. It received the Esri Special Achievement in GIS Award in 2010 for the Department's vision, leadership, hard work, and innovative use of Esri's GIS technology. St. Johns County was selected from more than 350,000 organizations worldwide.

Note: Information contained in this case study was provided through an interview with a representative from the GIS Division at the St. Johns County Department of Public Works.

WASHINGTON STATE DEPARTMENT OF TRANSPORTATION

Background

Washington State Department of Transportation's (WSDOT's) 7,000 employees oversee 18,500 lane-miles of highway and 3,600 bridges. The geography and the climate vary significantly between the moist and mild marine region on the west, through the Cascade Mountains in the center, to the dry, flat region in the east. WSDOT has a diverse fleet of vehicles and regionally dispersed management in place to maintain roadways in regions with very different temperature, precipitation, topographies, and population density. Portions of the Cascades can receive anywhere between 4 and 30 feet of snow during a typical winter, presenting a particular challenge to workers in the mountainous regions.

A GIS-based Maintenance Operations System to Monitor Activities and Performance

WSDOT's "Maintenance Operations" (formerly "Winter Operations") system collects real-time data from WSDOT maintenance vehicles across the state and displays the information on a browser-based map interface. The system enables employees from WSDOT's six geographic regions, who are responsible for maintaining regional roads, and the WSDOT headquarters Maintenance Operations group, who are responsible for overseeing the regions and budgeting, to manage plow operations and road conditions from a map. For any state road, staff can view up-to-date surface condition data and maintenance activity such as the material type being used for deicing and the number of plows being deployed to an area. The system also allows staff to watch a playback loop of historic data over a selected time period and develop reports of the data to debrief weather response activities and support operational decision-making.

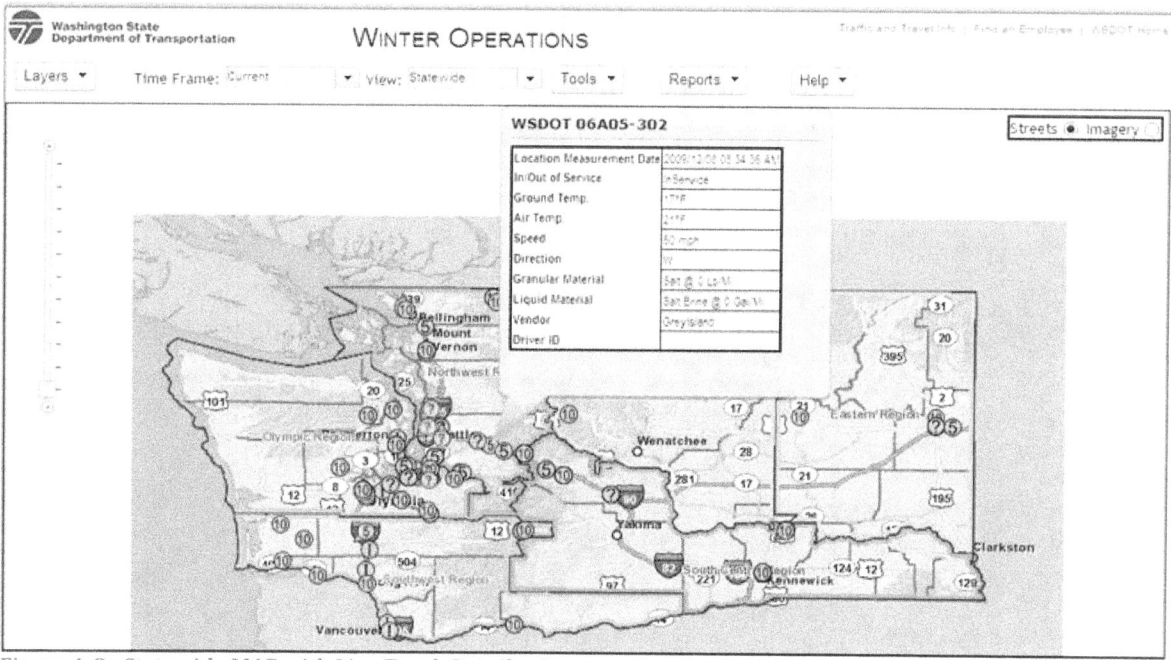

Figure A-2: Statewide MAP with Live Truck Details. Source: Alan Smith, WSDOT.

The application was originally built to integrate data from several separate vendor-hosted snowplow monitoring systems so that WSDOT staff could view plows across the state and capture data about resource allocation, materials usage, and weather conditions. Historically, each region had the autonomy to select equipment suppliers for snowplows. Over time, the state acquired equipment from four hardware manufacturers, several different software vendors (to control the spreaders and sprayers, communicate with sensors, and track material usage), and three real-time data viewers. Each system provided a different user interface and not all of them offered maps. Eventually, the WSDOT headquarters maintenance group asked the WSDOT GIS Branch to create an integrated GIS mapping application to view all plows and archive data. The current production system provides real-time interactive map for use throughout the agency, allowing improved asset management and more efficient use of existing resources.

The application displays asset-related data on a multi-functional, map-based interface. Data is collected from 250 plow trucks outfitted with sensors that automatically collect a wide range of information regarding the truck's activities and environmental conditions. Trucks are able to capture road and air temperature, material output rates, vehicle speed, direction of travel, driver identification (ID) number, and plow and spreader activity. Each event record is time stamped, GPS location stamped, and assembled into a profile of events ("points") by individual software vendors and then transmitted to WSDOT. Data is sent in near real time through a cellular connection or uploaded through a wireless connection at the maintenance facility if the vehicle has poor cellular reception during a trip. All data is managed by WSDOT in-house after it is downloaded. The raw data points are spatially associated to the nearest LRS segment by WSDOT. Event attributes are assigned to the associated line segment (road) for cartographic display and linear analysis. The data are stored in a Microsoft SQL server. An ArcGIS Server provides the geoprocessing and cartographic map service publication.

WSDOT employees can view the data on a browser-based map, zoom to the appropriate level of detail, and control the display of trucks and road segments, color-coded to indicate conditions (ice, snow, bare and wet, bare and dry, slush, etc.), and material used on the roadway (solid chemical, liquid chemical, sand, or other). Users can select a current "live" view or at specified point in time. Clicking on a truck or a state route line segment launches a pop-up window that provides detailed information about that feature. Last year, the GIS Branch enhanced the system further to make historical data available. Using an ArcGIS module, it added the functionality for users to view a playback of past conditions. Users select a start and end date and time range of interest and then click a "play". The map plays the track activities and road conditions for that timeframe while continuing to provide interactive map navigation.

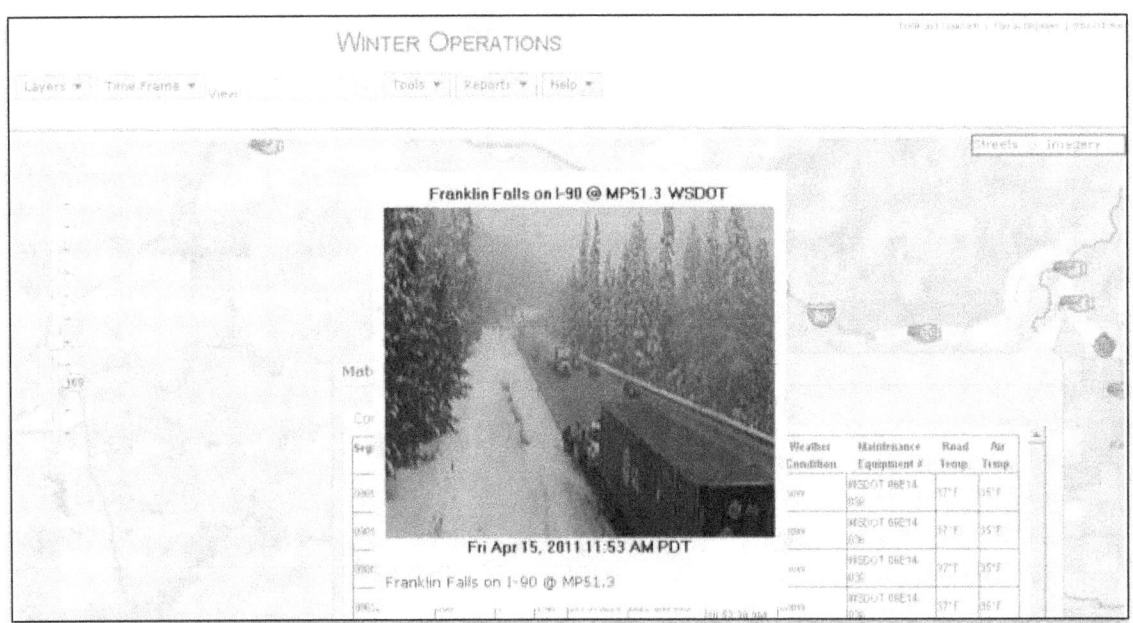

A-3: Live Materials Applied and Mapped with Road Conditions. Source: Alan Smith, WSDOT.

Benefits

Maintenance Operations has improved the ability of WSDOT staff to carry out basic asset, maintenance, and fleet management functions. The most common use of the system is to support staff with shift changes in the winter – incoming employees can see where trucks have been and where work still needs to be done. The color-coded map view of all state roadways also allows maintenance managers to quickly identify hazardous conditions and intuitively assess the geographic extent and relative severity. They can then manage plow routes and deploy trucks across regions as needed to address statewide conditions, localized weather events, and weather forecasts. Finally, WSDOT management can use the application to replay and debrief storm response activities for internal learning and in response to questions from the state transportation secretary or other officials.

The system has also provided some unexpected benefits to WSDOT departments outside of maintenance. WSDOT can now provide data-based responses to accident and litigation claims. WSDOT now has an accurate record of service level performance and a more timely understanding of liability. Traffic operators use the map interface as a situational awareness tool. The map includes WSDOT Incident Response vehicles, accident information, construction project locations and traffic camera feeds.

The services based architecture makes it very easy to add additional map content and makes all map service content available to other client applications.

Challenges

The quality of data and level of services that are collected from different vendors is inconsistent; some vendors are much more advanced than others. The data is homogenized into a single database, but the attributes vary depending on the capabilities of the vendor.

Future Plans

In the future, WSDOT employees see opportunities to use the data collected from Maintenance Operations to automate materials purchases and evaluate treatment methods. WSDOT is constantly purchasing roadway treatment material (e.g., salt, sand, deicers). Maintenance Operations allows the maintenance staff to systematically track materials usage to 97 percent accuracy, which is close enough for them to accurately monitor stock levels and automate materials requests when they run low. The system should also theoretically allow them to measure the effectiveness of different treatments supplied by various vendors on various road conditions. Right now, they have base data for two years that is not enough to make any conclusive statements but is providing a foundation for such analysis in the future.

WSDOT is also looking to expand Maintenance Operations to monitor other equipment such as mowers, line painters, and vacuum trucks. The maintenance group has already tried to apply the technology to paint striping trucks with mixed results. Overhead tree canopies interfere with the wireless signal needed to transmit the data back to the system. Also, the paint stripe location data needs to be accurate within 6 to 8 inches in order for the system to determine the directionality of traffic. The GPS equipment used to collect location information is somewhat inaccurate; and to compound matters, the spray nozzle is located at the end of a boom with a variable length that leaves two options: (1) placing the antenna at the end of the boom, which is risky and messy due to the proximity to paint; or (2) calculating an offset (which is complicated because it can vary). WSDOT employees have not yet come up with a feasible solution that provides accurate data.

Right now, the maps are available within the WSDOT intranet for internal use only and are not available to the public. There are some potential safety, security, and political issues with making all of the data public, but WSDOT is potentially interested in disseminating road condition information.

Note: Information contained in this case study was provided through an interview with a representative from the GIS and Roadway Data Office at the Washington State Department of Transportation.

COLORADO DEPARTMENT OF TRANSPORTATION

Background

Colorado Department of Transportation (CDOT) is responsible for a 9,146 mile highway system, including 3,447 bridges. Each year, this system handles over 27.4 billion vehicle miles of travel[8]. The agency has been emphasizing transportation asset management since the early 1990s, has one of the Nation's most mature pavement management systems, and was an early user of the Pontis bridge management system.

CDOT has six Engineering Regions, each being a semi-autonomous entity that is responsible for engineering, maintenance, planning and environmental, traffic, right-of-way and surveying, utilities, and human resource management for its area. Each region is also responsible for collecting its own asset data. The CDOT GIS program has been in existence since the mid-1980s and is centrally located in the GIS/Data Management Section of the Division of Transportation Development (DTD). By using a centralized GIS section, CDOT is striving to collect data in a more consistent manner among regions to not only be able to compare performance metrics uniformly but to also collect information that can be used for decision-making statewide among regions.

In 2006, CDOT implemented an Enterprise Resource Planning (ERP) system to streamline financial and human resource functions. CDOT selected SAP as its ERP system to replace, consolidate, and integrate numerous legacy and stand-alone, custom-developed software systems. In addition to redefining the business and information needs of CDOT, the ERP system also included components such as project scheduling and asset management and materials management. An important element of the ERP solution was to integrate geospatial technology within SAP to support the various business processes. GIS has been integrated with SAP to support modules for project scheduling and asset management as well as for streamlining reporting in accordance with FHWA requirements. SAP also keeps track of where funding for assets is coming from (revenue), and it monitors CDOT's spending on assets and the depreciation of assets.

Implementing GIS as a Project Management Tool

At CDOT, the integration of GIS with SAP is currently being used as a project management tool. A Web Mapping application interfaces with SAP, enabling CDOT's project and corridor managers to create, modify, and review information about a project's location and other features and attributes associated with that location. These features and attributes include structures on and jurisdiction information such as Counties, Transportation Planning Regions (TPR), Metropolitan Planning Organizations (MPO), Commission, and Congressional Districts. This project location information is updated in SAP databases and with GIS Spatial datasets, which are used in other web applications. Business and financial workflow related to the project can also be tracked within the Project Builder module. Projects from the STIP and CDOT long-range planning projects (LRP) are included in the system. ITS assets are managed in SAP, as is Maintenance and Operations (M&O), although M&O currently needs to be updated. There is an effort to put environmental and traffic operations assets in SAP.

Data Inventory - Collection, Integration, and Accuracy

CDOT recognizes that data integration and collecting data in a uniform manner plays a critical role in managing assets. In 2005, DTD initiated a policy to develop a single LRS for the entire Department so that any location-referenced data collected on all routes will be consistent and easily integrated with other corridor data. The single LRS was able to unify many different LRSs that had developed over the years between three business units and six regions into one linear reference framework for all CDOT highways. In the past, regions were not required to report changes to the inventory and often acted autonomously from headquarters.

---

[8] Colorado Department of Transportation, "About CDOT." Website. Last updated 23 Sep 2011. www.coloradodot.info/about

Implementing an LRS policy was particularly important for the integration of data into SAP as the unified LRS effort produced matching data segmentation across all highways for project development, field construction, safety, FHWA reporting, TAM, and planning. The project pushed DTD into updating the geometric alignments of the roadways in the GIS system and has greatly enhanced the geometric accuracy of the data. Each route was compared using beginning and end reference points and length in the three different databases. It was then matched against historical highway logs and evaluated using current data such as aerial imagery, digital video logs from the pavement management system, and as-built drawings provided by the regions, among others. Using a single inventory and common data source for all business units, SAP will enable consistent financial planning and management.

The unified LRS is now being used as the base layer for enterprise data collection. All routes in Colorado have been completed. Two years ago, CDOT began completely retracking its highway assets for inventory purposes using the single LRS. Some condition data have been collected as well. CDOT continues to update the Maintenance and Operations (M&O) data set, the Traffic Safety and Accident History data set, environmental assets, and transportation assets to reflect the new referencing system.

Tools for Data Collection

CDOT has developed an application called Storm Water Inventory Tool (SWIT) that uses Esri ArcGIS mobile and operates on a touchscreen tablet with a recreational Garmin GPS device. It is a numeric/alpha coding system designed to automate and standardize surveying and aerial mapping. A code is assigned when any topographic, drainage, utility, or aerial survey data is collected. The data is then exported into a standard GPS data format. CDOT has found that using a $100 recreational device can run the application as well as a $2,000 professional grade GPS device, enabling it to deploy and order devices quickly and easily. Currently, there are 12 devices operating in the field collecting environmental assets.

CDOT has started outsourcing more data collection and also wants to explore the possibility of collecting more asset data from the same collection activity. CDOT hopes to leverage existing processes for non-condition-related data collections where a CDOT staff member can remain at his desk and view assets from a video log that was collected by a contractor and gather the same information as if he were riding along the highway. Collecting asset data in this manner is potentially more cost effective than sending someone out into the field, and there is also increased safety in not having employees collecting data along a highway.

ITS Inventory and Maintenance Management System in SAP

The CDOT ITS Branch originally developed the ITS Maintenance Management System (ITS MMS) framework and detailed system functional requirements for another system. When CDOT implemented the SAP system, the ITS MMS functional requirements were migrated into SAP. CDOT's ITS inventory information is now completely housed in SAP. Using the inventory, ITS device as-built system plans are also being developed. Staff are able to perform maintenance activities using a work order-based system. CDOT tracks expenditures for labor, materials, equipment, warranties, device-life cycles, preventative maintenance protocols and cycles, and device and system condition.

GIS as a Visualization Tool

Much of the GIS data is available for download to the public through the CDOT website. MapView2 is an ArcGIS server application that provides mapping, query, route measures, and basic analysis tools. Users can produce customized maps from over 100 available layers. There is also a public-facing geographic web-based application that contains detailed information about the 2035 Statewide Transportation Plan corridors and STIP projects called ProLo. An interactive map is available for users to search and locate corridors and projects throughout Colorado.

Challenges

- Policy and Procedures: CDOT does not have a policy regarding data collection standards. There are also no procedures that require regions to work through a centralized location like DTD when they collect and use data. Many CDOT regions are developing data collection tools and standards without communicating or coordinating activities.
- Data Collection Coordination: DTD employees are now working with procurement to coordinate equipment purchases to standardize data collection initiatives. This will leverage an existing business process that will enable coordination and increase return on investment for CDOT.
- Staff limitations: Although there are 16 people in the GIS group, there are no true "data collectors" in the Section. The few data collection staff at CDOT are fully allocated toward traffic data collection. For example, when ITS Maintenance wanted to automate the system, it outsourced data collection in order to get the work done despite a limited CDOT staff.

Note: Information contained in this case study was provided through an interview with a representative from the GIS Data Management Section at the Colorado Department of Transportation.

MICHIGAN DEPARTMENT OF TRANSPORTATION

Michigan Department of Transportation (MDOT) is considered a national forerunner and leader in TAM. Nearly 60 professionals in an TAM Division at the MDOT headquarters in Lansing conduct TAM education and maintain the databases that house transportation data for the entire State of Michigan, including those owned by other public entities such as cities and counties. The TAM Division is also the champion and overseer of activities integrating TAM and GIS.

GIS for Public Reporting

MDOT is required to report annually on its TAM program to the State Legislature. GIS-based graphics are developed to show the location of highway, airport, and transit assets and its conditions for these annual reports. Maps are also developed for published reports to the public, such as the "Transportation System Performance Report" shown below. MDOT staff has found these map-based graphics to be effective in communicating the need for investment in the transportation system.

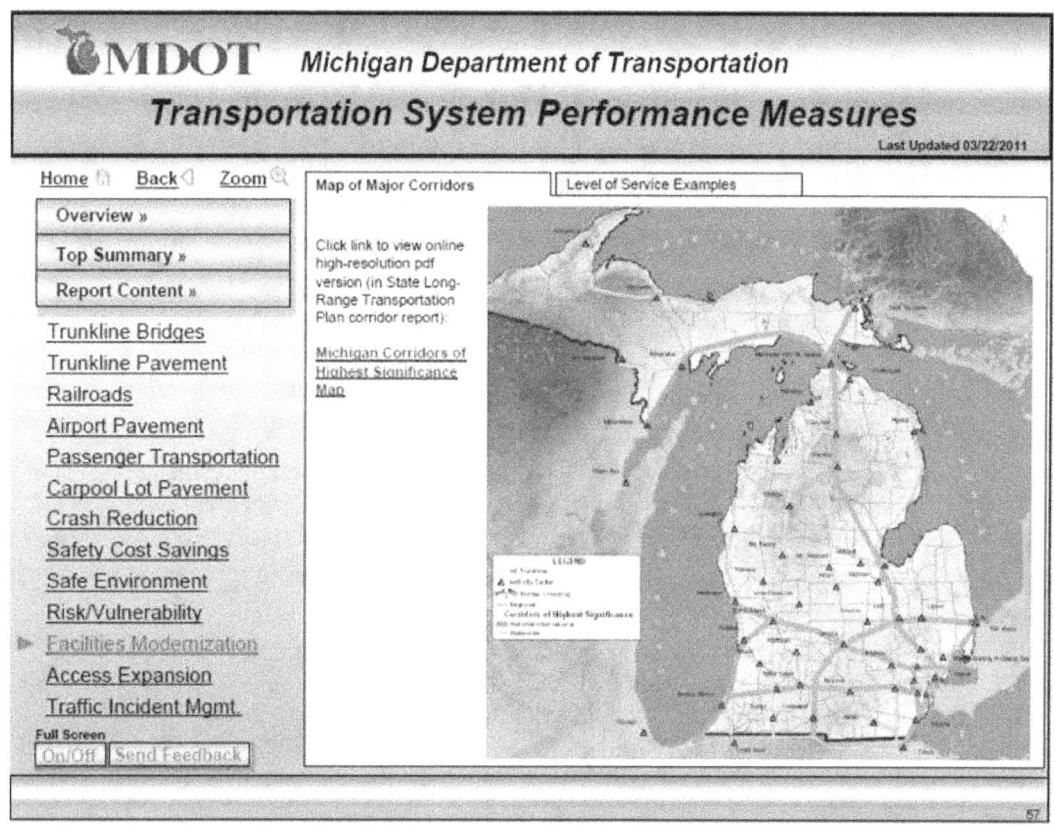

Figure A-4: Screenshot of MDOT's "Transportation System Performance Report." Source: MDOT website.

MDOT also provides asset inventory and condition data to the Michigan Transportation TAM Council (TAMC), a body of representatives from road operators established in 2007 to develop a coordinated TAM strategy for the state. The TAMC maintains an interactive map through which members of the public can view pavement and bridge condition ratings for all state-funded roads and bridges (shown below and accessible at http://tamc.mcgi.state.mi.us/MITRP/Data/paserMap.aspx).

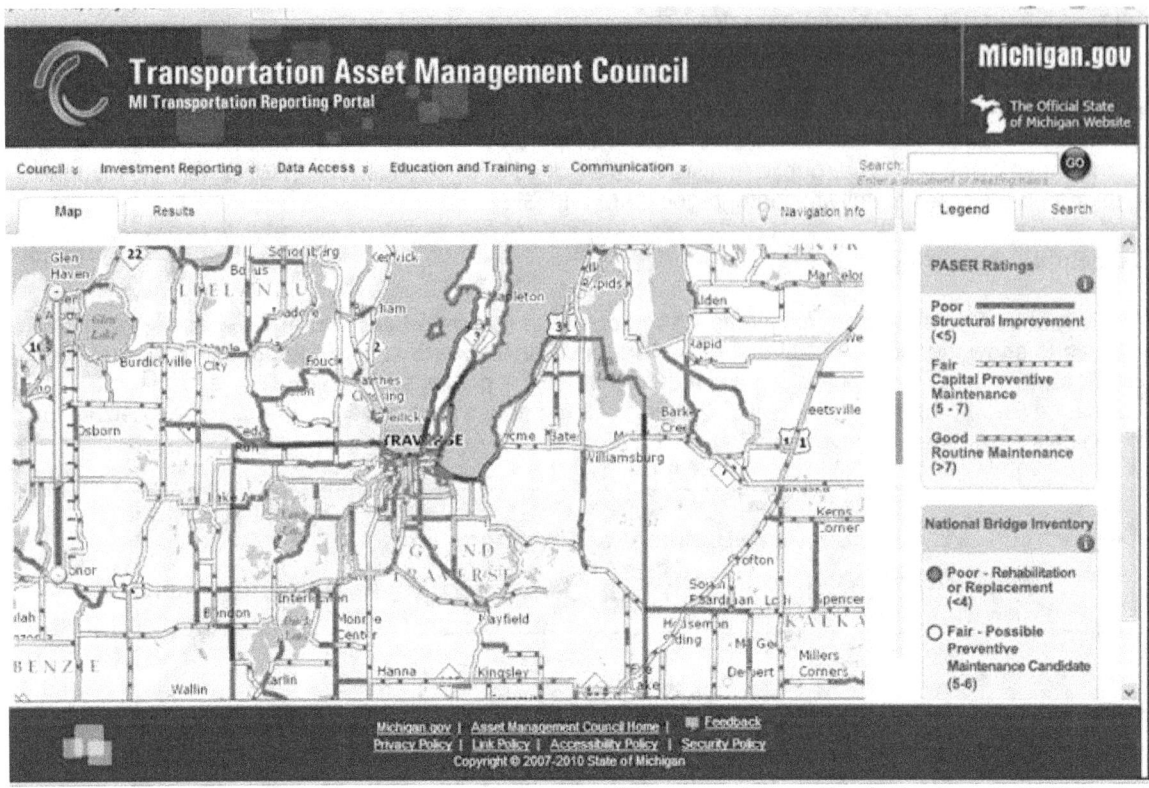

Figure A-5: Screenshot of Michigan's Transportation Asset Management Council website. Source: TAMC website.

GIS for Planning

Geospatial data is also used by MDOT staff for project planning purposes. GIS is the primary means through which staff can see the asset inventory and spatially analyze the condition of key assets. Staff in the Southwest Region Office in Kalamazoo use GIS-based TAM data to make budget decisions every year and assess the condition of roadside assets during project design phases.

A Diverse System Architecture

The TAM Division's GIS data is stored in a SQL (Esri) database hosted by the state Information Technology (IT) Department and can be accessed by MDOT headquarters and regional staff via the MDOT intranet. Base GIS maps are developed by the Center for Shared Solutions at the State of Michigan, a group that maintains base GIS networks for all departments in the state. The costs are shared among all participating departments, and one MDOT staffer is dedicated to updating this.

Transportation-specific data in the GIS database maintained by MDOT includes location and condition information for pavement, bridges, sign structures, guardrails, and catch basins. Condition data for bridges and pavement collected during inspections is stored in bridge and pavement management systems that are considered the authoritative sources. Data is pushed from these management systems to the Transportation Management System (TMS), an Oracle-based data warehouse and decision-support tool that was developed in-house to integrate data from several legacy systems and make it available over the internet (http://mdotwas1.mdot.state.mi.us/public/tms/). TMS then provides the bridge and pavement data to the GIS database. The GIS database (ArcGIS) is considered the authoritative source of data for other types of assets even if alternate forms of data are stored in other databases. The GIS database is used to update TMS annually for those assets.

A Wide Range of Staff Skills

Use of GIS varies widely from region to region. MDOT direct-force employees in the Southwest Region Office use Esri ArcGIS Server and ArcGIS viewing software to view and manipulate the data. They update it regularly after day-to-day maintenance work and use it annually make condition assessments and budget decisions. MDOT has also developed a data maintenance solution for the region. MDOT's smaller and less populated regions have limited staff with many job responsibilities and little time to use GIS software. Contract employees (or agencies) provide additional support in remote regions but tend also to be limited in their interest and knowledge in GIS. The disparity between regions places a burden on the headquarters staff to maintain advanced resources for the Southwest Region while also supporting the regions that cannot keep up. The headquarters must also train the regions on the use of GIS, which becomes a low priority when staff is limited, interest varies, and budgets are tight.

Data Standards in Flux

On the systems side, uncertainty over data standards and the potential for new, more integrated data management systems have slowed integration of GIS and TAM. The Department hopes to migrate all geodata to one system (e.g., into TMS from the existing asset database) and is currently debating whether Oracle Spatial Data or Esri standards make the most sense for the department. Moreover, the department aims to replace legacy TAM systems with enterprise systems but has yet to identify budget and leadership for doing so. MMS procurement was cancelled last year because of budget cuts, and a new sign management system (required by Federal mandate) with the potential for future expansion is the closest the agency can come to a new enterprise system.

MDOT employees recognize that these issues can be overcome with strong leadership that is willing to make investments of time and money in the program.

What's Next?

MDOT is developing a data business plan to address data standards issues and lay out an implementation plan for standardizing MDOT data.

MDOT has a bid out for enterprise resource management software. The immediate application is for sign management in response to the Federal reflectivity requirement, but MDOT employees recognize that the system could provide a basis for more enterprise TAM efforts In the future. It may also influence the data standards discussions.

Note: Information contained in this case study was provided through an interview with a representative from the Asset Management Division at the Michigan Department of Transportation.

OREGON DEPARTMENT OF TRANSPORTATION

Background

For many years, the Oregon Department of Transportation's (ODOT) asset management program consisted of several non-integrated programs. These programs were grouped by traditional areas of expertise such as pavement management, bridges, congestion, and safety. In 2005, ODOT proposed a pilot project to examine the many different types of existing data that were being collected and managed. The hope was to identify areas where the data overlapped or was related and areas where there were significant data gaps. As part of the pilot study, ODOT developed a basic inventory of asset condition data throughout the State to determine if the asset data was in fact up to date and accurate.

The pilot project, which spanned more than one year, produced insights and experiences that would ultimately help to redefine ODOT's asset management plan. Recognizing that the agency already had data in many different formats housed in many different places - and that new data was constantly being created - ODOT sought to reduce redundant data collection by making much of this data easily accessible to all ODOT staff through a web-based mapping tool called FACS-STIP. The acronym, FACS-STIP, stands for Features, Attributes, and Conditions Survey - Statewide Transportation Improvement Program.

While ODOT has practiced asset management efforts for several years; it officially created an Asset Management Integration Section in 2007. ODOT has what was formerly known as ITIS and Features Inventory, (now TransInfo) and several disparate systems, as well as more mature systems for bridge, pavement, and ITS. The agency is in the process of moving to a more enterprise system which is supported by asset management efforts.

FACS-STIP TOOL

An internal tool to ODOT, FACS-STIP, seeks to integrate GIS with TAM. In FACS-STIP, GIS helps to combine efforts via one user-oriented, enterprise tool. ODOT staff all over the state often needs information about some feature of the highway. This can be for scoping possible construction projects or just research for various work tasks. The intent of the FACS-STIP is to make asset attribute facts available and support decisions related to preservation or improvement projects. In creating the FACS-STIP Tool, ODOT acknowledged that sound management of asset data – including spatial data and related asset condition information – is essential for operations and maintenance, project delivery and construction activities, as well as strategic transportation planning. All efforts are extremely important to a state's overall transportation program; and, with the right approach, the data could be effectively shared through GIS.

Data

Today, the FACS-STIP Tool database is housed in SQL server 2008 with ArcGIS running in the foreground. The configuration allows staff to access all statewide data for mapping and spatial analysis while leaving database operations and maintenance to the technical staff in ODOT's IT department. Reports generated from the FACS-STIP Tool are designed primarily for in-house use, and ODOT does not have any immediate plans to make the FACS-STIP Tool available to the general public.

Much of ODOT's GIS data comes from the many unique county transportation agencies across the State. Basemap data often originates with the emergency 911 call centers, but each county's data collection and maintenance business processes were developed before the State was able to organize. Currently, the State is working on developing a standard data exchange format that will enable all 36 Oregon counties to streamline data collection processes and facilitate easy transfer of basemap data to the State. Efforts are currently focused on road centerlines, and the goal is to make consistent, statewide basemap data available to the general public via download.

ODOT hopes to use the FACS-STIP Tool to facilitate sharing data across the agency, but asset data collection processes are often defined by the business processes of each specialized department. Many of these processes have been in place for years. Based on the subject matter – pavement, unstable slopes, project tracking, right-of-way boundaries, for example – the data often stand alone and are managed and maintained independently. The FACS-STIP Tool seeks to integrate all these unique datasets so that, in addition to their individual functions, they can also be used in conjunction with each other.

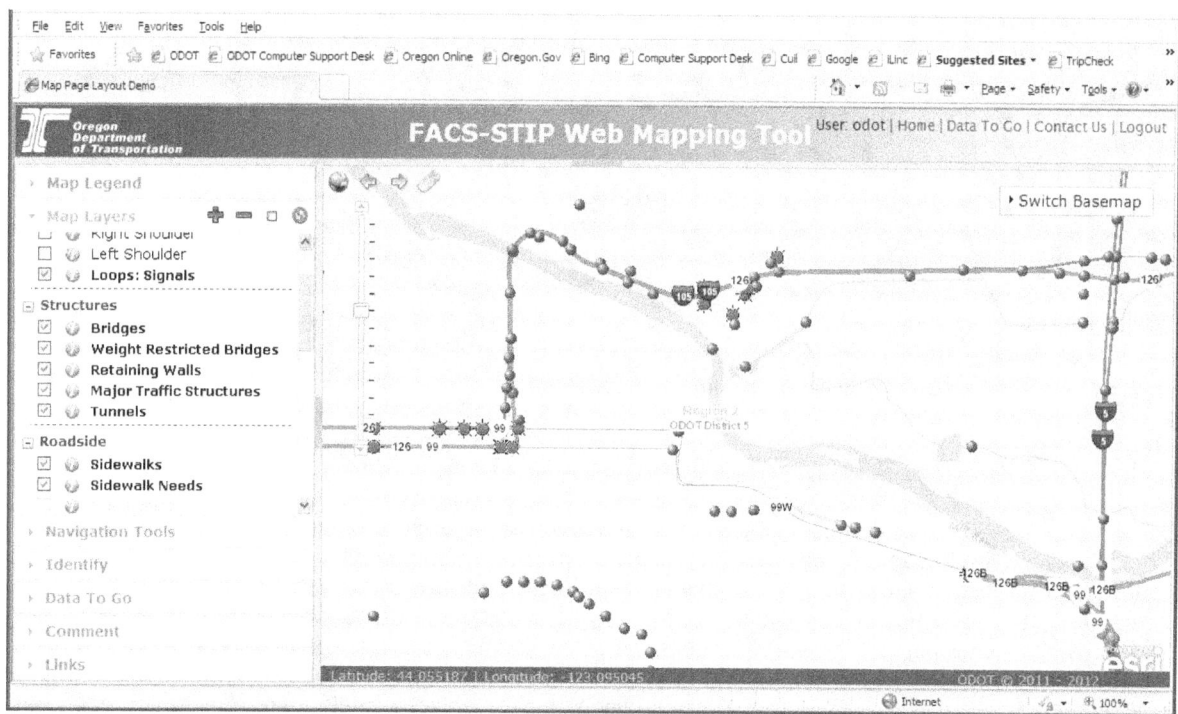

Figure A-6: Screenshot of FACS-STIP's "Asset Visibility Controls," for map layer selections. Source: ODOT

The FACS-STIP Tool uses a simple user interface with one toolbar with a dropdown menu of tools (see figure above). From this menu, users can control map data visibility by selecting any combination of basemap layers (i.e. roads, orthoimagery, planning districts, political boundaries), transportation data layers (i.e., traffic models, crash data, highway width), and project layers (i.e., bridge maintenance, expansion, pavement rehabilitation).

A unique feature of the FACS-STIP Tool is the Map Commenting feature that allows users to select any point or line on the map and insert a comment for other users to view. Users can also click on a link to the RSS feed to see who authored the comment, what region they are from, and the type of comment (project related, asset related). Related documents (such as spreadsheets or contracts) can be uploaded and attached to the comment as well.

ODOT has also a tool called Data2Go as part of the FACS-STIP Tool that is used in conjunction with the tool's web mapping capabilities (see figure following). It allows the user to query, investigate, and export data for a particular area of interest using the FACS-STIP's Web Map interface. Asset reports can be viewed on screen or exported to Excel as necessary. The Data2Go tool provides a platform for sharing data visually as many users are more easily able to recognize data needs when they are shown on a map.

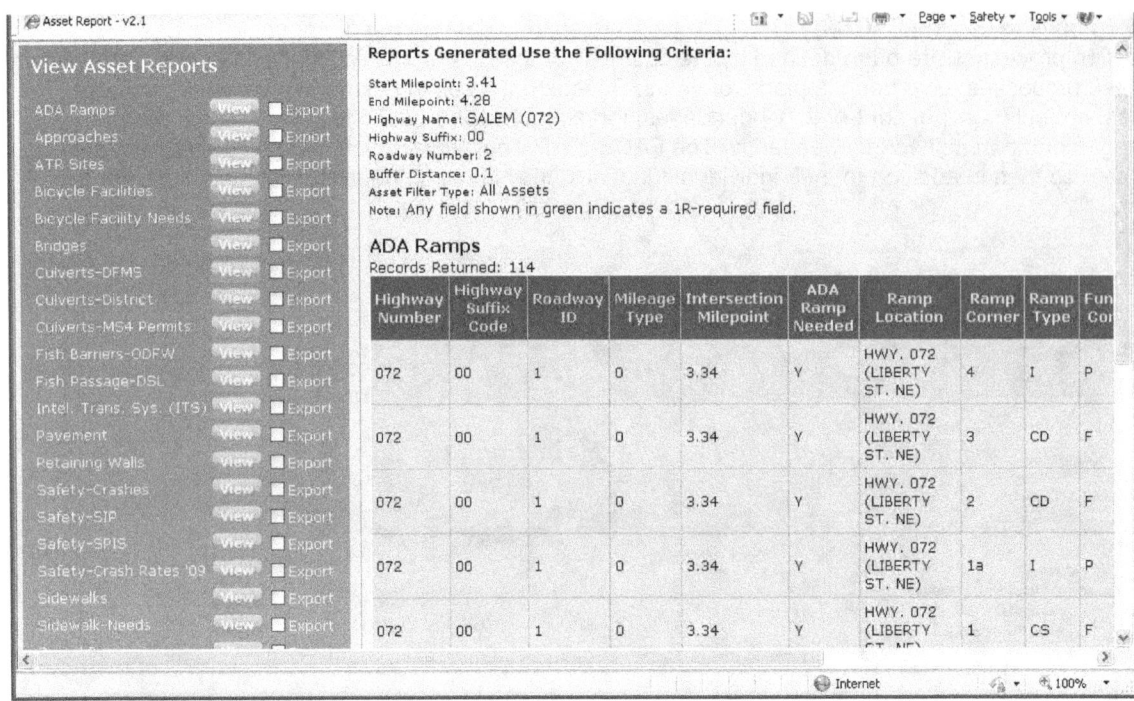

Figure A-7: Sample asset reports viewed on screen using FACS-STIP Tool. Source: ODOT.

Since the FACS-STIP's original launch, ODOT has periodically incorporated additional features or functionality related to common practices and procedures that would benefit from inclusion in the application. For instance, the agency has always been required to periodically complete roadside inventories, but programatic permissions created new specific requirements for certain projects. FACS-STIP provided a new platform on which to base the roadside inventory, and the department took advantage of the opportunity to populate the tool with inventory data from disparate sources and manage it in a centralized location. Unique business needs sometimes conflict with standard operating procedures and can create challenges to integration of linear highway asset data. Location, location, location has been the ODOT mantra that allows the central GIS unit to convert data to similar formats. This allows ODOT to then integrate and share this data.

Given that asset data had been traditionally gathered and managed in several different areas of ODOT, the agency embarked on a significant outreach effort throughout the development of FACS-STIP Tool. They asked each specialized department which assets and what data were most important. With input from the experts, decided which data to include in FACS-STIP and developed a Beta version of the tool. The department then demonstrated the tool across the State in order to end up with functionality and design that would meet many needs. In 2008, the first version of the FACS-STIP Tool officially went live and was generally well received by the many tool users across the agency. The agency found that buy-in from its IT department was especially crucial to FACS-STIP's success.

FACS-STIP is regarded as a successful, ongoing initiative, and ODOT is supportive of the tool's continual development through funding. However, department resources are limited, expertise can be difficult to come by, and the tool does suffer from growing pains from time to time. The old business "silo" model is generally seen as detrimental to the tool's goal, yet breaking down the silos and integrating business processes takes a persistent, steady effort and must be based on good standards..

Other Applications

TransGIS 2.0
Just released in autumn 2011, TransGIS will continue to be developed into a public facing mapping tool designed for users of many skill levels. Incorporating much of the data already collected and housed by

FACS-STIP, the tool offers an interactive format that allows users to manipulate an on-screen map by turning layers on and off.

Digital Video Log
The digital video log (DVL) is another tool made available by ODOT's Road Inventory and Classification Services Unit. The DVL provides digital images of the state highway system – taken every 100th of a mile – in both increasing and decreasing mile directions and includes a straight-ahead view and a side view at 45 degrees.

Mobile GPS Tools
ODOT is building a suite of mobile GPS tools that operate somewhat independently. Efforts around environmental data management have led the way, but work currently in process will create additional applications for asset field data collection. It is hoped that these will ultimately be integrated with the the FACS-STIP Tool to streamline data collection and updates. Currently, the tool is used to export data for roadside inventory purposes.

FME (Feature Manipulation Engine) – GIS data warehouse
ODOT developed a statewide data warehouse that employs Safe Software's FME tool to transform spatial data (features) from different software systems into like formats that can be viewed, manipulated, and/or analyzed with the same program. FME aims to streamline data conversion processes that can be extremely repetitive and time consuming.

earthmine
An interactive mapping system, called *earthmine*, was recently tested by ODOT. The product tested was imagery with a point cloud behind it. Plug-in tools allowed ODOT staff to "collect" asset data at their desks. Data was successfully collected for most of the fourteen (14) different assets that were included in the pilot. A total of one hundred miles of highways were included, but it remains an option to expand beyond the pilot.

Note: Information contained in this case study was provided through an interview with representatives from the Geographic Information Services Unit, Transportation Data Section and the Asset Management Integration Section, Technical Services Division at the Oregon Department of Transportation.

OHIO DEPARTMENT OF TRANSPORTATION

Background

The Ohio Department of Transportation currently operates the seventh largest highway system in the United States and the sixth largest interstate system measured by total lane-miles. These highways support the fifth greatest traffic volume by total vehicle miles and contain the second largest inventory of bridges in the nation, totaling over 43,000. Ohio DOT maintains approximately 49,000 lane miles of highway system statewide.

In recent years, the Ohio Department of Transportation has been working on implementing a comprehensive TAM system. In 2009, the State's Transportation Asset Management Committee put together a list of recommendations to help move Ohio's asset management processes forward. The Ohio DOT Office of Technical Services was charged with recommending a framework that would allow for the establishment of a centralized asset inventory database to support the Department's decision making processes. Having the data stored geo-spatially allows for easier integration of all data that can be conveniently mapped and exchanged.

Within the Department, data typically resides in individual databases specific to particular asset classes – for example, pavement data is housed in a different database than maintenance data. In the past, the needs for each department were determined on an individual basis without regard to integration. The Department hopes that by integrating the information together, decisions, resource allocation, and coordination can occur as one system. Also, with a centralized database, consistent information can be distributed both horizontally and vertically through the organization.

Enterprise Web-Based Geographic Information System Application

Ohio DOT is currently working on developing an enterprise web-based geographic information system application ("WebGIS") to more effectively distribute and communicate information related to transportation assets. The application will allow users to:

- Display map views of assets and their attributes;
- Pan, zoom, and measure distances;
- Turn layers on and off;
- Perform advanced data queries; and,
- Easily update data layers as needed.

Integration of data sources is a key functionality of WebGIS. The application will employ user-friendly technology and integrate with various Ohio DOT existing systems, such as Ohio DOT's linear referencing system (the Base Transportation Referencing System (BTRS) which consolidates the department's various referencing systems) and existing applications, such as the video-log viewer application (PathWeb) and the straight-line diagram web application. Google Streetview will also be incorporated with online basemap layers, including aerial imagery, topography, terrain, and shaded relief.

Innovative Asset Management Tools Using GIS

Ohio DOT has many new ideas about where to take its GIS and asset management programs. Rolling out new products over the entire state can be difficult, expensive, and risky. To better manage its resources and program outcomes, the department looks to its districts, counties, and municipalities to develop and test ideas and products before adopting a new program statewide. Districts have been developing customized in-house applications for the State. Ohio DOT District 2 has been the forerunner in developing several asset management applications. Ohio DOT District 1 has assisted with the implementation of these technologies.

Mobile Video Log
District 2 is currently working on applications in advanced remote asset collection, such as the ODOT Video Log that uses a van to collect data and projects the GPS coordinates into the image as it is collected. After evaluating several vendors with video log solutions for asset extraction, the DOT determined that all were too expensive and none fit their requirements exactly. This resulted in the District building its own asset capture software. The software was developed for scalability, flexibility, and independence from any vendor. IT was also able to leverage existing resources and investments. The video log software was used to capture most of the districts' sign inventory. In one summer using two interns to collect data, District 2 was able to collect data for approximately 40,000 signs, averaging 200 – 400 signs per day.

Geotagging Photographs
Ohio DOT District 2 geotags photographs taken with its digital camera (see Figure A-8) to make innovative use of photography in its GIS software. The district has made a business decision to replace all of its end of life digital cameras with GPS enabled cameras. With automated process these images are discoverable over the WAN and displayed on a map. By mapping these images, both geographically and temporally, will allow asset managers, or anyone else in the organization, a way to view images taken from other departments at areas and times that may be of interest to them. These cameras also record the direction the person is facing while taking the picture. Saving the sensor data allows a level of documentation that was previously difficult to achieve and cost prohibitive.

Figure A-8: Ohio DOT District 2 geotagged photograph displayed on a map. Source: GSAM.

Mobile Culvert / Asset Application
Ohio DOT has also developed GPS software for asset collection. This software is currently being implemented statewide for culvert management. Central Office identified it as a best practice and standardized it for the rest of the 12 districts using the software developed by District 2 as well as the District's database structure. There are 48,000 culverts in the inventory and with an estimated additional 100,000 to be added by the end of the data conflation process. Initially the software was developed and deployed by District 2 in 2008, then standardized in 2009 by District 1. The culvert application is now fully implemented in districts 1,2,4,5,6,7,8 and 12. The project is headed by the Office of Hydraulic

Engineering. Recently, Ohio DOT collected over 68,000 barrier assets within a timeframe of 3 weeks utilizing 25 college interns due to the merging of GIS technology and asset management tools.

Utilizing Unmanned Aircraft System (UAS) Technology for Remote Sensing
Ohio DOT will be testing the potential uses of unmanned aircraft, or drones, which have become more commercially affordable. The DOT recently purchased a SenseFly Swinglet CAM, a mini-drone with a high-resolution camera, which is set to fly next spring. The flight plan of a drone can be pre-programmed; and while video is the main form of data that can be collected at this time, there is the potential for extensive automated data collection. The Swinglet CAM will produce on-demand high resolution aerial imagery. The cost of collecting aerial imagery data is significantly less expensive than with a traditional aircraft. Ohio DOT is also working with the Federal Aviation Administration (FAA) on all aircraft and flying requirements.

Decision-Making

Although Ohio DOT has been managing assets for many years, the recent integration of mapping technology allows the agency to display assets visually. While the agency does not often work with spatial analysis, the ability to simply make asset maps has proven quite successful as it pertains to influencing decision makers and agency executives. According to the Department, a simple color-coded map can make a strong impact on those in decision-making positions. Ohio DOT staff remarked that it is not always easy to get buy-in from leadership to invest in the technology, but executives and decision-makers have enjoyed the results, seeing technical data presented on easy-to-read maps.

Note: Information contained in this case study was provided through an interview with representatives from the Division of Planning, Office of Technical Services and the District Two Office at the Ohio Department of Transportation.

APPENDIX B: GLOSSARY AND ACRONYMS

3-D	Three-dimensional	Imagery that provides a sense of depth as well and height and length.
AASHTO	American Association of State Highway and Transportation Officials	Trade organization for transportation agencies and advocacy group for transportation-related issues.
Architecture, system		The design of the database, processes, and technologies that make up an information system.
ArcView		Desktop GIS software developed by ESRI used for basic GIS operations and print maps.
CMMS	Computerized Maintenance Management System	Maintenance Management System used by St. John's County, Florida.
COTS	Commercial Off-the-Shelf	Professional software or tools purchased commercially and used with minimal development costs.
CDOT	Colorado Department of Transportation	For more information: http://www.coloradodot.info/
Digitize		The process of assigning digital coordinates by physically or automatically tracing hard copy documents. Used for converting paper maps, aerial photos, or raster images into digital form.
DOT	Department of Transportation	Public agency responsible for operating and maintaining transportation infrastructure in a geographic area.
DynSeg	Dynamic Segmentation	A GIS feature that allows tabular data to be associated with static geographic features within GIS based on a common linear referencing scheme rather than be mapped independently
ESRI	Environmental Systems Research Institute	ESRI is the company that makes ArcInfo, ArcView, and many other related software for GIS.
FACS-STIP	Features, Attributes, Conditions Survey - Statewide Transportation Improvement Program	Oregon DOT's internal web-based mapping tool that gives all staff access to data about STIP projects.
Feature		A spatial element which represent a real-world entity by having specific characteristics. Often used synonymously with the term object. A generalized description of a point, line or polygon.
Geocoding		The process by which the geographic coordinates of a location are determined by its address, postal code, or other explicitly non-geographic descriptor.

Geotag		Adding geographic identification information to data (e.g., geotagging photographs adds location coordinates to a photo)
GIS	Geographic Information System	A system that allows users to analyze, manage, and view geographically (geo)-referenced data on a map. GIS provides the ability to query data, report, and perform statistical analysis while also providing the benefit of visualization and geographic analysis that is uniquely offered by maps.
GIS-T	GIS for Transportation Symposium	AASHTO-sponsored annual opportunity for both public and private industry interested in the use of GIS for transportation purposes to collaborate and share experiences
GPS	Global Positioning System	Satellite-based navigations system that allows people to identify specific locations.
GSAM	GIS for Strategic Asset Management Group	An informal group of state transportation agency staff focusing on issues that GIS professionals face with the collection and management of large asset inventories.
LRS	Linear Referencing System	Reference system for numbering and locating data along linear assets, such as roads.
ODOT	Oregon Department of Transportation	For more information: http://www.oregon.gov/ODOT/
ProLo	Project Locator	Colorado DOT's interactive web application that allows the public to see STIP projects on a map.
Query		A way of selecting features based on a set of common characteristics. For example, the act of selecting all the building that have an area greater than 2000 sq. ft. out of a database.
Remote sensing		The process of obtaining information about land, water, or an object without any physical contact between the instruments doing the sensing and the subject. Remote sensing most often refers to collecting data using instruments aboard aircraft or satellites.
RFIP	Road Features Inventory Program	WSDOT GIS-enabled application for recording information about equipment along roadsides.
SAP		A software corporation that makes enterprise software to manage business operations and customer relations. The SAP ERP application is an integrated enterprise resource planning (ERP) software used by Colorado DOT to boost information consistency and accuracy, improve and automate operations, and save support costs.
TAM	Transportation Asset	The systematic collection, storage, and analysis of

	Management	data related to physical property in order to identify maintenance and construction needs, plan for the future, and strategically allocate limited resources.
WSDOT	Washington State Department of Transportation	For more information: http://www.wsdot.wa.gov/

APPENDIX C: INTERVIEW PARTICIPANTS

Agency	Department / Section	Contact
Colorado Department of Transportation	GIS / Data Management Section	
Michigan Department of Transportation	Asset Management Division	MDOT-assetmgt@michigan.gov
Ohio Department of Transportation	Division of Planning, Office of Technical Services	
Ohio Department of Transportation	District Two Office	Fred.Judson@dot.state.oh.us (419) 373-4497
Oregon Department of Transportation	Geographic Information Services Unit, Transportation Data Section	odot.maps@odot.state.or.us 503-986-3154
Oregon Department of Transportation	Asset Management Integration Section, Technical Services Division	503-986-4092
St. Johns County, Florida, Department of Public Works	Department of Public Works, GIS Division	(904) 209-0266
Washington State Department of Transportation	GIS and Roadway Data Office	TDOAdmin@wsdot.wa.gov 360-570-2350